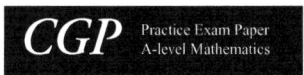

Edexcel A-Level Mathematics

Practice Set 1
Paper 1: Pure Mathematics 1

Time allowed: 2 hours

Centre name					
Centre number					
Candidate number					

Surname	
Other names	
Candidate signature	

In addition to this paper you should have:
- An Edexcel mathematical formula booklet
- A calculator

Instructions to candidates
- Use black ink or ball-point pen.
- A pencil may be used for diagrams, sketches and graphs.
- Write your name and other details in the spaces provided above.
- Answer all questions in the spaces provided.
- Show clearly how you worked out your answers.
- Round answers to 3 significant figures unless otherwise stated.

Information for candidates
- There are 16 questions in this paper.
- There are 100 marks available for this paper.
- The marks available are given in brackets at the end of each question.
- You may get marks for method, even if your answer is incorrect.

Advice to candidates
- Work steadily through the paper and try to answer every question.
- Don't spend too long on one question.
- If you have time at the end, go back and check your answers.

For examiner's use			
Q	Mark	Q	Mark
1		9	
2		10	
3		11	
4		12	
5		13	
6		14	
7		15	
8		16	
Total			

Exam Set MEP71

© CGP 2018 — copying more than 5% of this paper is not permitted

Answer ALL the questions.

Write your answers in the spaces provided.

1 a) Given that f(x) = $x^3 - 4x^2 - 3x + 7$, find f'(x).

(2)

b) Hence find the values of x for which f(x) is a decreasing function, giving your answer in the form $\{x : x > a\} \cap \{x : x < b\}$ where a and b are real numbers to be found.

(3)

2 A helicopter flies between 3 locations, A, B and C, which are positioned such that AB = 9 km, AC = 5 km and angle ABC = 24°.
Find the possible values of angle ACB to 1 decimal place.

(3)

3 a) Express $\left(\sqrt{2}\,x^{\frac{1}{3}}\right)^6$ in the form ax^b where a and b are integers to be found. **(2)**

b) Hence find the x-coordinates of the points where the line $y = 1 - 3x$ intersects the curve with equation $\left(\sqrt{2}\,x^{\frac{1}{3}}\right)^6 + 6x - 18 = y^2 - y$. **(4)**

4 For each of the following, prove that the statement is false.

a) The exterior angles of a regular n-sided polygon are always acute. **(1)**

b) For $n \in \mathbb{R}$, $n \neq -1$, $\dfrac{n}{n+1} \geq 0$. **(1)**

c) For $n < 50$, if n is an odd prime then one or both of $n + 2$ and $n + 4$ are prime. **(1)**

5 The elastic energy stored in a large industrial spring, E, in joules, is directly proportional to the square of how far it is extended, x, in metres.

When the spring is extended by $(\sqrt{2} - 1)$ m, it has $(7\sqrt{2} - 9)$ joules of elastic energy.

Find the exact amount of elastic energy in the spring when it is extended by $\sqrt{2}$ m, giving your answer in joules in the form $a + b\sqrt{2}$, where a and b are integers. **(5)**

6 The circle C has equation $x^2 - 8x + y^2 + 4y - 29 = 0$.
The centre of C is at the point X.

a) Find:

i) the coordinates of the point X. **(2)**

ii) the radius of the circle C. **(1)**

A tangent from the point $P(-16, 13)$ touches the circle at the point Y.

b) Find the distance PY. (3)

7 a) Express $\dfrac{4x^2 + 8x + 3}{4x^2 - 9}$ as a single fraction in its simplest form. (2)

b) Hence, or otherwise, solve the equation:

$$\log_3(4x^2 + 8x + 3) - \log_3(4x^2 - 9) = 2, \quad x > 1.5$$

(4)

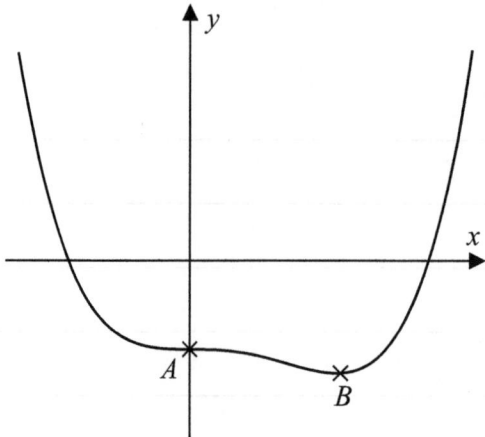

Figure 1

Figure 1 shows the graph of $y = f(x)$ where $x \in \mathbb{R}$. The graph has stationary points A and B.

a) State the nature of the stationary point A, justifying your answer with reference to the shape of the graph. **(2)**

b) Explain why f(x) does not have an inverse function. **(1)**

B is a minimum turning point with coordinates (p, q), where p and q are constants.

c) Write down, in terms of p and q, the coordinates of the point B under these transformations:

i) $y = f(x - 1)$ **(1)**

ii) $y = 3f(2x)$ **(1)**

d) Given that f(x) = $3x^4 - 2x^3 - 2$:

  i) find the exact values of *p* and *q*.

  (4)

  ii) justify that *B* is a minimum turning point.

  (2)

e) Sketch the graph of *y* = |f(−*x*)|.

  (2)

9 Figure 2 shows how a manufacturer cuts pieces of cheese to sell.

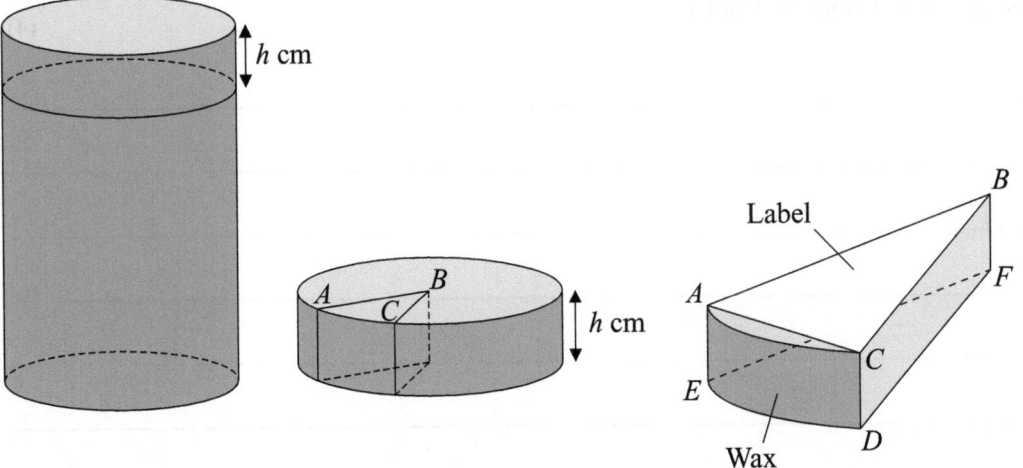

Figure 2

A cylinder of cheese has a layer of wax with negligible thickness applied to its curved surface.

The cylinder is sliced horizontally, h cm below the top face.
The slice is then cut vertically along two radii, AB and CB, as shown above.
Each piece, $ABCDEF$, has a triangular label, ABC, applied to the top face.

For a particular piece, $h = 4$ cm, angle $ABC = 1.02$ radians and the area of the label is 85.9 cm². Find the area of wax on this piece of cheese.

(4)

10 a) The deduction is only valid if f(x) is continuous on the interval [a, b]. If f has a discontinuity (e.g. an asymptote) between a and b, then the sign change does not guarantee a root exists between a and b.

b) i)
- Error 1: The student used $x_1 = 3.9$ (a rounded value) in the calculation of x_2 instead of the more accurate value $x_1 = 3.9308176...$
- Error 2: The student stopped after only two iterations — they have not iterated enough to be confident the answer is correct to 3 d.p. (successive approximations have not yet agreed to the required accuracy).

ii) Using $x_0 = 5$:

$x_1 = 5 - \dfrac{2(5)^3 + 9(5) - 125}{6(5)^2 + 9} = 5 - \dfrac{170}{159} = 3.9308176...$

$x_2 = 3.9308176 - \dfrac{2(3.9308176)^3 + 9(3.9308176) - 125}{6(3.9308176)^2 + 9} = 3.6176567...$

$x_3 = 3.6176567 - \dfrac{f(3.6176567)}{f'(3.6176567)} = 3.5918812...$

$x_4 = 3.5917770...$

So $\alpha = 3.592$ (3 d.p.)

11 A company expects to make a profit of £250 000 in the year 2021.
The company's yearly profit is then expected to increase by 5% per year.

a) By forming an inequality and solving it algebraically, work out which year will be the first to have an expected profit of more than £500 000.
(4)

b) Find the total expected profit for the company from 2021 to 2030 (inclusive), giving your answer to the nearest hundred pounds.
(2)

12 a) Express $\dfrac{18x-7}{(2-3x)(1+x)}$ in partial fractions. (3)

The first three terms in the binomial series expansion of $\dfrac{1}{1+x}$ in ascending powers of x are:
$$1 - x + x^2, \quad |x| < 1$$

b) Hence find the first three terms, in ascending powers of x, of the binomial series expansion of $\dfrac{18x-7}{(2-3x)(1+x)}$. (5)

13 Prove that:

$$\frac{\sin 2\theta + \sin \theta}{\cos 2\theta + 1 + \cos \theta} = \tan \theta \qquad \cos \theta \neq 0, -\frac{1}{2}$$

(4)

14 a) Use a suitable substitution to show that:

$$\int \frac{75e^{-x}}{(1 + 3e^{-x})^2} \, dx = \frac{k}{1 + 3e^{-x}} + C$$

where k is an integer to be determined.

(5)

A population of kangaroos is being studied. The population is modelled by the differential equation:

$$\frac{dP}{dt} = \frac{75e^{-t}}{2P(1+3e^{-t})^2} \quad t \geq 0$$

where P is the population of kangaroos in thousands, and t is the time measured in years since the study began. There were 5500 kangaroos when the study began.

b) Solve the differential equation, giving your answer in the form $P^2 = f(t)$. **(4)**

c) Find the limit of the size of the population as $t \to \infty$, showing clear algebraic working. **(2)**

15 A factory, which makes metal cubes, contains a machine that heats the cubes to strengthen them. When heated, the volume of a cube increases at a constant rate of 1.5 cm^{3}s^{-1}.

If the rate of increase of the total surface area of a cube is higher than 1 cm^{2}s^{-1}, the cube will be damaged.

Showing your working clearly, find the minimum side length that a cube must have so that it doesn't become damaged when heated.

(5)

16 Figure 3 shows a design for the cross-section of a man-made river bank.

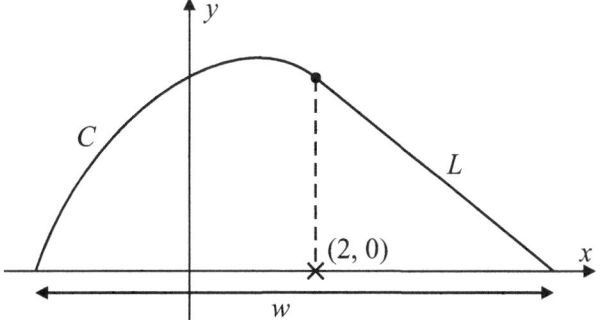

Figure 3

The design is bounded by the curve C, the line L and the x-axis and has a width of w.
Curve C has equation $3x^2 - 2xy + 2y^2 = 18$ where $x \leq 2$ and $y \geq 0$.
Line L is the tangent to the curve C at the point where $x = 2$.

a) Find the coordinates of the point where line L meets curve C. (5)

b) Find the width of the design, w. (5)

END OF QUESTIONS

10. Figure 3 shows a design for the cross-section of a man-made river bank.

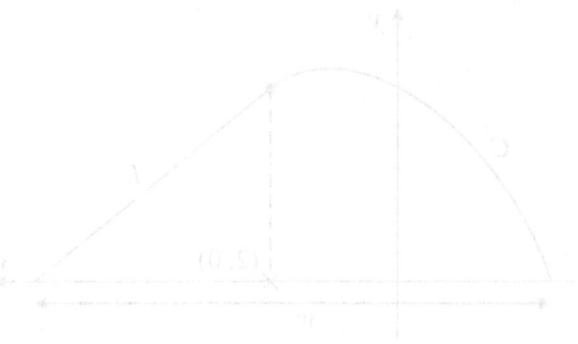

Figure 3

The design is bounded by the curve C, the line L and the x-axis and has a width of w.
Curve C has equation $4x^2 2xy + 2y^4 = 18$ where $x \leq 2$ and $y \geq 0$.
Line L is the tangent to the curve C at the point where $x = 2$.

a) Find the coordinates of the point where line L meets curve C.

(5)

b) Find the width of the design, w.

(5)

END OF QUESTIONS

Edexcel A-Level Mathematics

Practice Set 1
Paper 2: Pure Mathematics 2

Time allowed: 2 hours

Centre name					
Centre number					
Candidate number					

Surname
Other names
Candidate signature

In addition to this paper you should have:
- An Edexcel mathematical formula booklet
- A calculator

Instructions to candidates
- Use black ink or ball-point pen.
- A pencil may be used for diagrams, sketches and graphs.
- Write your name and other details in the spaces provided above.
- Answer all questions in the spaces provided.
- Show clearly how you worked out your answers.
- Round answers to 3 significant figures unless otherwise stated.

Information for candidates
- There are 16 questions in this paper.
- There are 100 marks available for this paper.
- The marks available are given in brackets at the end of each question.
- You may get marks for method, even if your answer is incorrect.

Advice to candidates
- Work steadily through the paper and try to answer every question.
- Don't spend too long on one question.
- If you have time at the end, go back and check your answers.

For examiner's use			
Q	Mark	Q	Mark
1		9	
2		10	
3		11	
4		12	
5		13	
6		14	
7		15	
8		16	
Total			

Exam Set MEP71

Answer ALL the questions.

Write your answers in the spaces provided.

1   In the expansion of $(2 + px)^5$, the coefficient of x is 240 and the coefficient of x^2 is q.

   a)   Write down the constant term in the expansion of $(2 + px)^5$. (1)

   b)   Find the value of p. (2)

   c)   Find the value of q. (2)

2   Two fishing boats are stationary at points A and B. Point A has a position vector of $(4\mathbf{i} + 9\mathbf{j})$ km, and point B has a position vector of $(12\mathbf{i} - 3\mathbf{j})$ km, both relative to a fixed origin O.

   An inspection boat moves directly from the point X, with position vector $(4\mathbf{j})$ km, to a point Y on the line AB. The distance from point Y to the boat at B is exactly three times its distance from point Y to the boat at A.

   a)   Find $\overrightarrow{XY}$ in its simplest form. (3)

b) Find the distance the inspection boat moves, giving your answer in km as a simplified surd. **(2)**

3 An economist uses the following model to predict how many subscribers, S, in millions, a social media website will have t years from today:

$$S = 18 + 7t - t^2 \quad \{t : t \geq 0\} \cap \{t : t \leq 9\}$$

a) Explain why the values of t in the model are restricted to those given above. **(2)**

b) Find the rate of change, with respect to time, in the number of subscribers predicted by the model 6 months from today. **(2)**

c) Write S in the form $a - (t + b)^2$, where a and b are rational numbers to be found. **(2)**

d) Hence state the maximum number of subscribers predicted by the model. **(1)**

4 a) Given that f(x) = $x^3 + x^2 - 8x - 12$ and f(-2) = 0, factorise f(x) completely.
(4)

b) Sketch the graph of y = f(x), stating the coordinates of the points where the graph intersects the x- and y-axes.
(3)

c) Given g(z) = $e^{6z} + e^{4z} - 8e^{2z} - 12$, use your answer to part a) to find the exact solution of the equation g(z) = 0.
(2)

5 a) Show that the equation $2x^2 + kx - 2 = 0$, where k is a real constant, has real solutions for all values of k. **(2)**

b) Show that:
$$k\tan\theta - 2\cos\theta = 0, \quad \cos\theta \neq 0$$
can be written as $2\sin^2\theta + k\sin\theta - 2 = 0$. **(3)**

c) Hence solve $3\tan\theta - 2\cos\theta = 0$ for $0° < \theta < 360°$. **(3)**

6 A manager of a chemical company models the volume of a gas, V, in cubic metres, as being inversely proportional to its pressure, P, in Newtons per square metre.

Some of this gas is stored in a cylindrical tank of radius 2 m. For safety reasons, the gas is to be transferred to another cylindrical tank of the same height.

a) Find the radius that this new tank will have to be so that the pressure of the gas is reduced by 84%.

(4)

b) The manager claims that, whenever gas is transferred to a cylinder with double the radius and the same height, the pressure is decreased by 75%. Prove that the manager is correct.

(2)

7 A sequence is defined by:

$$u_{n+1} = pu_n - 3, \text{ where } u_2 = 45 \text{ and } u_3 = 33.$$

a) Find the exact value of $\sum_{r=1}^{4} u_r$. **(3)**

b) The sequence converges to a limit L as n tends to infinity. Show algebraically that $L = -15$. **(2)**

8 a) Use the trapezium rule with 3 strips to find an approximate value of:

$$\int_{0.4}^{1} \tan^2 x \, dx$$

where x is measured in radians.

(4)

b) Use your answer to part a) to find an approximate value of $\int_{0.4}^{1} (\sec^2 x + 4) \, dx$.

(3)

9 Given that x is small, show that $\dfrac{\operatorname{cosec} 2x(1-\cos 4x)}{6x} \approx \dfrac{2}{3}$. (3)

10 a) Show that $\displaystyle\int_1^4 \ln x\, dx = \ln a + b$, where a and b are integers to be found. (5)

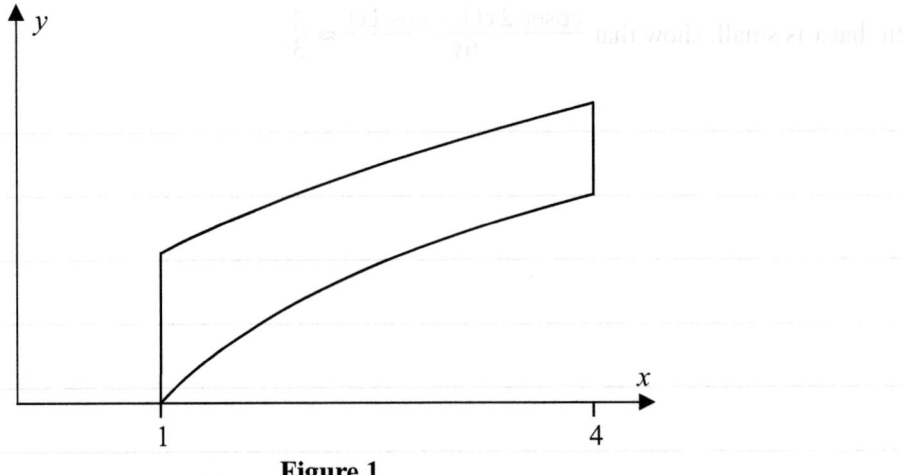

Figure 1

Figure 1 shows the cross-section of the roof of a bus shelter. Its area, A, is bounded by the curves $y = \sqrt{x}$ and $y = \ln x$ and the lines $x = 1$ and $x = 4$, with units in metres.

b) Find the exact value of A in m^2, giving your answer in the form $p - \ln q$, where p and q are constants.

(4)

11 Given that $f(\theta) = \cos\theta$ where θ is measured in radians, prove from first principles that $f'\left(\frac{\pi}{3}\right) = -\frac{\sqrt{3}}{2}$.

You may use the formula for $\cos(A + B)$ and the small angle approximations.

(6)

12 Find the equation of the normal to the curve:

$$y = \frac{5x^4 - 6}{3e^{2x}}$$

at the point with x-coordinate 0. Give your answer in the form $ax + by + c = 0$, where a, b and c are integers.

(6)

13 The rate at which a certain plant grows during spring is affected by the temperature, C, in degrees Celsius. The situation can be modelled using the following equations:

$$C = 20 - \frac{9}{t+1} \qquad \frac{dH}{dt} = \frac{1}{10}t^{0.2} \qquad 0 \leq t \leq 90$$

where H is the height of the plant in centimetres, and t is the number of days since the start of spring. Calculate $\frac{dH}{dC}$ at $t = 5$.

(4)

14 Use proof by contradiction to show that, for a right-angled triangle, the length of the hypotenuse is less than the sum of the lengths of the other two sides.

(4)

15 A geography student models the depth of water in a harbour, d, in metres, using the equation $d = \sqrt{3} \sin \frac{t}{2} - \cos \frac{t}{2} + 5$, where t is the number of hours from the start of the study.

a) Write down the depth of water in the harbour predicted by the model at the start of the study. **(1)**

b) Express $\sqrt{3} \sin \frac{t}{2} - \cos \frac{t}{2}$ in the form $R \sin\left(\frac{t}{2} - \alpha\right)$ where $0 < \alpha < \frac{\pi}{2}$. **(3)**

c) Hence find the minimum depth of water in the harbour predicted by the model, and the time to the nearest minute when this minimum first occurs. **(3)**

To add in a predicted sea level rise of 1.2×10^{-5} metres per day, the student refines the model so the equation becomes $d = \sqrt{3} \sin \frac{t}{2} - \cos \frac{t}{2} + 5 + kt$.

d) Find the value of k. **(1)**

16 Given that $f(x) = \dfrac{1}{(3x-1)^2}$:

a) state the value of x which must be excluded from the domain of f. **(1)**

$x = \dfrac{1}{3}$

Given that $g(x) = 2\cos x + \dfrac{1}{3}$, $x \in \mathbb{R}$:

b) state the range of g. **(2)**

$-\dfrac{5}{3} \leq g(x) \leq \dfrac{7}{3}$

c) Find, in its simplest form, $\displaystyle\int \left(fg(x) + 6f(x)\right) dx$. **(5)**

$fg(x) = \dfrac{1}{(3(2\cos x + \tfrac{1}{3}) - 1)^2} = \dfrac{1}{(6\cos x)^2} = \dfrac{\sec^2 x}{36}$

$6f(x) = \dfrac{6}{(3x-1)^2}$

$\displaystyle\int \left(\dfrac{\sec^2 x}{36} + \dfrac{6}{(3x-1)^2}\right) dx = \dfrac{\tan x}{36} - \dfrac{2}{3x-1} + C$

END OF QUESTIONS

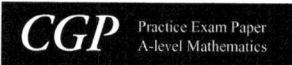

Edexcel A-Level Mathematics

Practice Set 1
Paper 3: Statistics and Mechanics

Time allowed: 2 hours

Centre name	
Centre number	
Candidate number	

Surname	
Other names	
Candidate signature	

In addition to this paper you should have:
- An Edexcel mathematical formula booklet
- A calculator

Instructions to candidates
- Use black ink or ball-point pen.
- A pencil may be used for diagrams, sketches and graphs.
- Write your name and other details in the spaces provided above.
- Answer all questions in the spaces provided.
- Show clearly how you worked out your answers.
- Round answers to 3 significant figures unless otherwise stated.

Information for candidates
- There are 13 questions in this paper.
- There are 100 marks available for this paper.
- The marks available are given in brackets at the end of each question.
- You may get marks for method, even if your answer is incorrect.

For examiner's use

Q	Mark	Q	Mark
1		8	
2		9	
3		10	
4		11	
5		12	
6		13	
7			
Total			

Advice to candidates
- Work steadily through the paper and try to answer every question.
- Don't spend too long on one question.
- If you have time at the end, go back and check your answers.

Exam Set MEP71

Answer ALL the questions.

Write your answers in the spaces provided.

SECTION A: STATISTICS

1. The random variable X is normally distributed with mean μ and standard deviation σ.

   a) State the value of a, if $P(X = \mu) = a$. **(1)**

   b) State the value of b, if $P(X \geq \mu) = b$. **(1)**

   c) Find the value of c, if $P(X < \mu + c\sigma) = 0.8$. **(2)**

2. Rajad is investigating the daily maximum wind gusts (measured in knots) at Heathrow in June 2015. He decides to sample 10 of the 30 readings from the month using systematic sampling.

   a) Describe briefly how he could obtain a suitable systematic sample. **(1)**

The ten readings in Rajad's sample are:

23 27 24 14 19 18 24 20 18 20

Rajad considers any reading which falls outside the interval given by mean ± (2.5 × standard deviation) to be an outlier.

b) Showing your working clearly, determine if there are any outliers in Rajad's sample. **(4)**

Letitia is also investigating the daily maximum wind gusts in knots at Heathrow. She finds the data for June 1987, and summarises it using a box plot, shown in Figure 1.

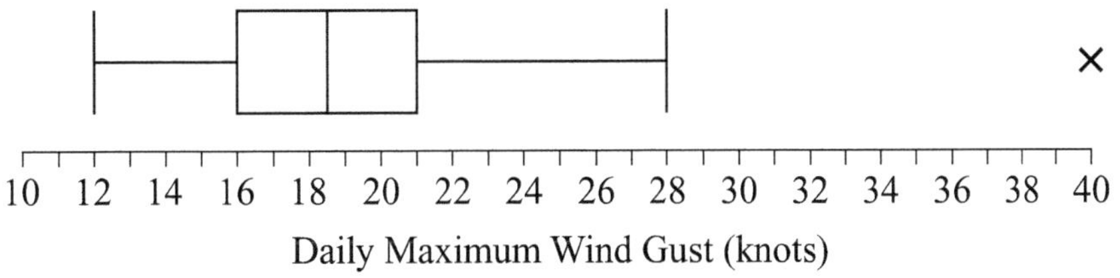

Figure 1

c) By calculating appropriate measures of central tendency and variation, compare the distributions of the daily maximum wind gusts at Heathrow throughout June 1987 and June 2015. **(4)**

Leave blank

Rajad wants to continue his investigation by looking into the daily maximum wind gusts in Camborne, and comparing the data to Heathrow.

d) Using your knowledge of the large data set, explain whether you would expect the daily maximum wind gusts to generally be higher or lower in Camborne than Heathrow. **(1)**

3 Simon plays golf at his local 18-hole golf course. In his last 20 visits, he works out that he has scored above par on 60% of the holes he has played. For each visit, he models the number of holes where he scores above par, H, using a binomial distribution: $H \sim B(18, 0.6)$.

a) Give one assumption that Simon has made in modelling H in this way. **(1)**

b) Each time Simon plays golf, he tests to see if the probability that he scores above par has changed. Stating your hypotheses clearly, find the critical region for such a test at the 10% significance level. **(5)**

4 Chloe is investigating the 24-hour rainfall total, t mm, and the daily mean pressure, P hPa, in Perth on several rainy days in July 2015. She takes a sample of the data where $t > 0$, and codes the data for pressure using the coding $c = P - 1000$.

Chloe calculates the equation of the linear regression line of t on c to be $t = 35 - 1.25c$.

a) Find, in its simplest form, the equation of the linear regression line of t on P. (1)

Chloe then calculates the product moment correlation coefficient of the data, and obtains a value of $r = -0.786$.

b) Give an interpretation of this value in context. (1)

Chloe thinks that an exponential model, of the form $t = kb^{(P-1000)}$, might be a better fit for the data. She draws a scatter diagram, plotting the values of $\ln t$ on the vertical axis against the values of $(P - 1000)$ on the horizontal axis. She then draws a line of best fit, which has a gradient of -0.2 and goes through the point $(0, 5.9)$.

c) Use this information to find, to 3 significant figures, appropriate values for k and b. (3)

5 A doctor tests 20 patients every weekday for a condition which is present in 5% of patients. The probability that a patient has the condition is independent from that of any other patient, and each patient is only tested once. On any weekday, work out the probability that the doctor tests:

a) more than 3 patients with the condition, (2)

b) at least 16 patients who do not have the condition. (2)

The doctor wants to estimate the probability that, over the next five weekdays, no more than 10 patients will have the condition. He approximates the number of patients with the condition over five weekdays, C, with a normal distribution.

c) Use a normal approximation to estimate the probability that C is no more than 10. (3)

d) Give a criticism of the doctor's use of a normal approximation in this situation. (1)

6 A and B are events such that $P(A) = 0.7$ and $P(B) = 0.15$. Find the value of:

a) $P(A \cap B)$ if A and B are mutually exclusive. (1)

b) $P(A \cup B)$ if A and B are independent. (2)

c) If A and B are dependent and not mutually exclusive, such that $P(A \cap B) = 0.037$, find:

i) $P(A' \cap B)$ (1)

ii) $P(A' | B')$ (2)

7 A random sample of the times taken by 120 athletes to complete a marathon in 2018 are summarised in the table below.

Time, t (minutes)	$130 \leq t < 135$	$135 \leq t < 140$	$140 \leq t < 145$	$145 \leq t < 150$	$150 \leq t < 170$
Number of athletes	20	32	37	21	10

a) Two of these athletes are picked at random.
Find the probability that exactly one of these athletes is in the $130 \leq t < 135$ class. (2)

b) When the data is represented on a histogram, the total area of the histogram is A cm² and the rectangle representing the $140 \leq t < 145$ class has a height of x cm. Work out the height and width of the rectangle representing the $150 \leq t < 170$ class in terms of x and A. **(3)**

The marathon takes place every year. Each year, the organisers of the marathon record the time taken by each runner to complete the race, T minutes. Using data from all previous marathons, they find that T follows a normal distribution, with mean 144 minutes, and standard deviation 7 minutes.

The organisers believe that, on average, runners were faster in 2018 than in previous years.

c) Estimate the mean of the sample from 2018, and hence test, at the 2.5% significance level, whether or not there is enough evidence to support the organisers' claim that the mean time taken was lower in 2018 than in previous years. State your hypotheses clearly.
You may assume that the standard deviation remained unchanged. **(6)**

SECTION B: MECHANICS

[Unless otherwise indicated, give your answers to either 2 or 3 significant figures, and take $g = 9.8$ ms^{-2}.]

8 A motorcyclist is taking part in a race on a straight course. The acceleration-time graph below shows his motion between the start of the race, and the point where he crosses the finish line.

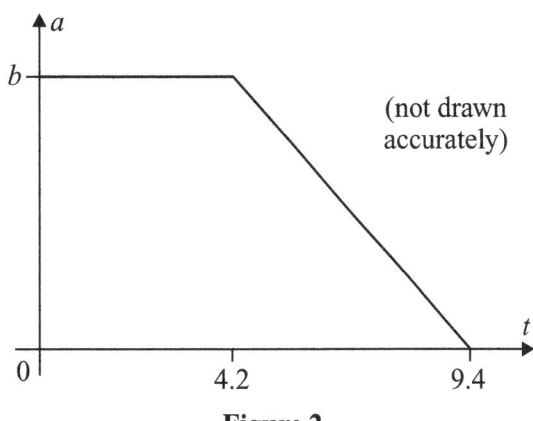

Figure 2

The motorcyclist is at rest on the start line and then accelerates at b ms^{-2} for 4.2 seconds.
After changing gear, the acceleration of the motorcyclist decreases at a constant rate, and reaches 0 ms^{-2} when he crosses the finish line at $t = 9.4$ seconds.
The velocity of the motorcyclist at the finish line is 57.8 ms^{-1}.

a) Show that the value of b is 8.5 ms^{-2}. (2)

The motorcyclist brakes immediately on reaching the finish line and decelerates at 12.5 ms^{-2} in a straight line. He continues to brake until his velocity is 10 ms^{-1}.

b) Find the time elapsed, and the distance travelled by the motorcyclist, between crossing the finish line and reaching 10 ms^{-1}. Give your answers to 2 significant figures. (4)

The motorcyclist was filmed by a drone during the race.
The displacement vector of the drone, **s**, in metres, is given by:

$$\mathbf{s} = \left(\left(\frac{t^3}{15}\right)\mathbf{i} + (t-2)\mathbf{j}\right) \text{ metres}, \quad 0 \leq t \leq 9.4$$

where t is the time in seconds after the motorcyclist started the race.

c) Find the speed of the drone at the time when the motorcyclist finished the race, giving your answer to 3 significant figures. (4)

d) Explain why it is not appropriate to use the kinematic (*suvat*) equations to model the motion of the drone. (1)

9   A 1250 kg car is towing a 750 kg trailer along a straight horizontal road. The trailer is connected to the car by an inextensible chain which is kept taut and parallel to the road. Constant resistance forces of magnitudes B N and $4B$ N are assumed to act on the car and trailer respectively.

When a driving force of 4900 N acts on the car, both car and trailer accelerate constantly at 1.5 ms^{-2}.

a) Find the value of B. (3)

b) Find the exact tension in the chain. **(2)**

Leave blank

c) Describe one assumption made about the chain which is not mentioned in the model. **(1)**

10 An object of weight 325 N, modelled as a particle, is at rest on a rough plane that is inclined at angle α to the horizontal, where $\tan \alpha = \frac{5}{12}$.

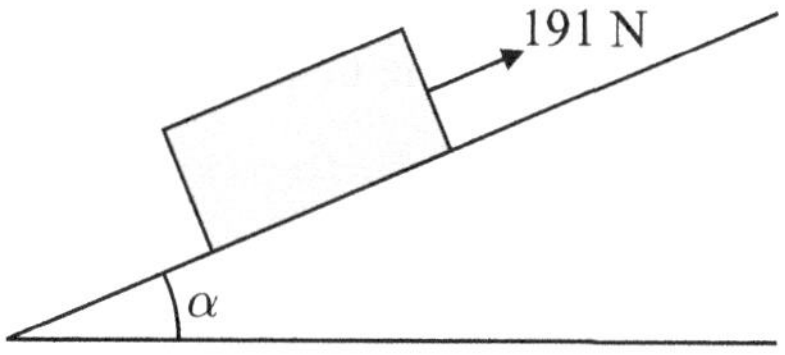

Figure 3

A force of 191 N is applied to the object parallel to the slope, as shown in Figure 3, so that the object is on the point of moving up the slope.
Find the coefficient of friction, μ, between the object and the slope. **(6)**

11 A footbridge across a river is built by placing a 6 m long plank, of mass 86 kg, horizontally on two supports at its end points, A and B.

A child of mass 52 kg stands on the plank at the point X so that the plank is in equilibrium and the magnitude of the reaction force at B is two thirds of the reaction force at A.

The child is modelled as a particle and the plank as a rigid uniform rod.

a) Show clearly that the magnitude of the reaction force at A is 82.8g N. **(2)**

b) Find the distance AX. Give your answer in cm to the nearest whole cm. **(3)**

c) Explain how you have used the modelling assumptions that:

  i) the child is a particle. **(1)**

  ii) the plank is rigid. **(1)**

12 *[In this question, take the value of g to be 9.81 ms⁻² and give your answers to 3 significant figures.]*

An aeroplane is ascending in a straight line over a lake, travelling at a constant speed of 225 ms⁻¹ at an angle of elevation of 3°. An aid package is released from the aeroplane, and travels a horizontal distance of 1.9 km before landing in the lake.

The package is modelled as a particle with initial velocity equal to the velocity of the aeroplane. Find:

a) the height of the aeroplane above the ground at the moment the package was released, (3)

b) the speed of the package at the moment just before it hit the water, (3)

c) the length of time that the package was further above the water than when it was released. (2)

13 *[In this question, **i** and **j** are horizontal unit vectors in the East and North directions respectively.]*

Ice hockey is played with a rubber disc called a puck. At the start of one ice hockey match, Marta hits the puck from the point with position vector $(30\mathbf{i} + 13\mathbf{j})$ m. The puck is modelled as a particle which slides horizontally on a level surface at a constant velocity of $(-4\mathbf{i} + 6\mathbf{j})$ ms^{-1}.

a) For the puck to have a constant velocity, state one assumption that has been made about the playing surface. **(1)**

b) Find the direction in which the puck is moving, giving your answer as a three-figure bearing. **(2)**

Another player, Cindy, starts the match at the point with position vector $(34\mathbf{i} + 22\mathbf{j})$ m. She starts at rest and immediately moves with a constant acceleration of $(-a\mathbf{i})$ ms^{-2}.

c) Given that she intercepts the puck, find the value of a. **(5)**

d) After Cindy intercepts the puck, she moves with it to the point *P* with position vector (15**i** + 15**j**) m, then shoots the puck along the ice at the goal with velocity (−44**i** − *A***j**) ms⁻¹.

Goalpost G_1 has a position vector of (4**i** + 12**j**) m.
Goalpost G_2 has a position vector of (4**i** + 14**j**) m.
Both are modelled as having negligible width.

The goaltender is stood on the goal line at point *Q* with position vector (4**i** + 12.5**j**) m.
She can block the puck up to 0.8 m from her position.

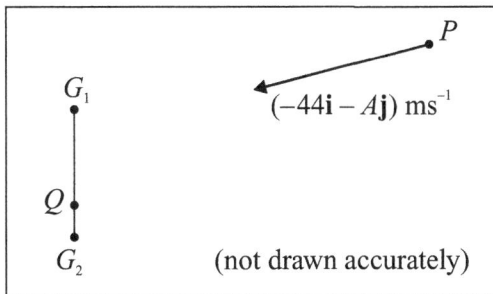

Figure 4

Calculate the range of possible values of *A* for which Cindy will score the goal. **(4)**

END OF QUESTIONS

d) After Cindy intercepts the puck, she moves with it to the point P with position vector $(15\mathbf{i}-15\mathbf{j})$ m, then shoots the puck along the ice at the goal with velocity $(-14\mathbf{i}-A\mathbf{j})$ m s^{-1}.

Goalpost G_1 has a position vector of $(\mathbf{i}+13\mathbf{j})$ m.
Goalpost G_2 has a position vector of $(4\mathbf{i}-14\mathbf{j})$ m.
Both are modelled as having negligible width.

The goaltender is stood on the goal line at point Q with position vector $(3\mathbf{i}+12.5\mathbf{j})$ m.
She can block the puck up to 0.6 m from her position.

Figure 4
(not drawn accurately)

Calculate the range of possible values of A for which Cindy will score the goal.
(4)

END OF QUESTIONS

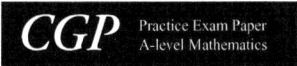

Edexcel A-Level Mathematics

Practice Set 2
Paper 1: Pure Mathematics 1

Time allowed: 2 hours

Centre name	
Centre number	
Candidate number	

Surname	
Other names	
Candidate signature	

In addition to this paper you should have:
- An Edexcel mathematical formula booklet
- A calculator

Instructions to candidates
- Use black ink or ball-point pen.
- A pencil may be used for diagrams, sketches and graphs.
- Write your name and other details in the spaces provided above.
- Answer all questions in the spaces provided.
- Show clearly how you worked out your answers.
- Round answers to 3 significant figures unless otherwise stated.

Information for candidates
- There are 15 questions in this paper.
- There are 100 marks available for this paper.
- The marks available are given in brackets at the end of each question.
- You may get marks for method, even if your answer is incorrect.

Advice to candidates
- Work steadily through the paper and try to answer every question.
- Don't spend too long on one question.
- If you have time at the end, go back and check your answers.

For examiner's use			
Q	Mark	Q	Mark
1		9	
2		10	
3		11	
4		12	
5		13	
6		14	
7		15	
8			
Total			

Exam Set MEP71

Answer ALL the questions.

Write your answers in the spaces provided.

1. Given that $f(x) = 3x^2 + 12x + 19$:

   a) find the value of the discriminant and explain what it tells you about the graph of $y = f(x)$. **(2)**

   b) write $f(x)$ in the form $p(x + q)^2 + r$, where p, q and r are integers to be found. **(3)**

   c) sketch the graph of $y = f(x)$, showing clearly the coordinates of any points of intersection with the axes and any stationary points. **(3)**

d) Describe a combination of two transformations that would map the graph of $y = 3x^2 + 12x + 19$ onto the graph $y = x^2$. (2)

2 The organisers of a car show want to rope off a rectangular area of floor space.
- They have 40 m of rope available.
- They need to have at least 60 m² of space roped off.
- The length of the rectangle must be exactly twice its width.

Find the difference between the maximum and minimum width of the rectangle. Give your answer in metres, in the form $\dfrac{a - \sqrt{b}}{c}$, where a, b and c are integers. (4)

3 a) Show that when $y = e^{2x}$, y is directly proportional to $\frac{dy}{dx}$. (2)

b) The line l is the tangent to the curve $y = e^{2x}$ at the point $(2, e^4)$.
Find, using exact values, the coordinates of the point where l crosses the x-axis. (2)

4 Freya wants to prove that $3\sin x + x^2 + 3 > 0$ for all values of x.

a) Freya makes the following claim, "Because $\sin x$ repeats every 2π, I only need to prove it for $0 \leq x \leq 2\pi$." Is she correct? Explain your answer. (2)

b) Prove that $3\sin x + x^2 + 3 > 0$ for all values of x. (3)

5 The graph of $y = f(x)$ intersects the x-axis at $x = -2$. Given that:

$$f(x) = x^3 - bx^2 + 2x + 40$$

find the value of b and express $f(x)$ as the product of 3 linear factors.

(6)

6 A rubber ball is dropped from a height of 6 m.
It falls vertically before hitting the ground and bouncing straight back up.
After the first time it hits the ground, it ascends to a height of 5.52 m.
The maximum heights that the ball reaches after each bounce form a geometric sequence.

    a) Show algebraically that the first time the ball bounces to a maximum
height of less than 1 metre is after the 22nd bounce. (5)

    b) Eventually the ball comes to rest. Estimate the total distance that the ball travelled. (2)

7 A square has vertices at (0, 4), (4, 0), (0, −4) and (−4, 0) as shown in Figure 1.

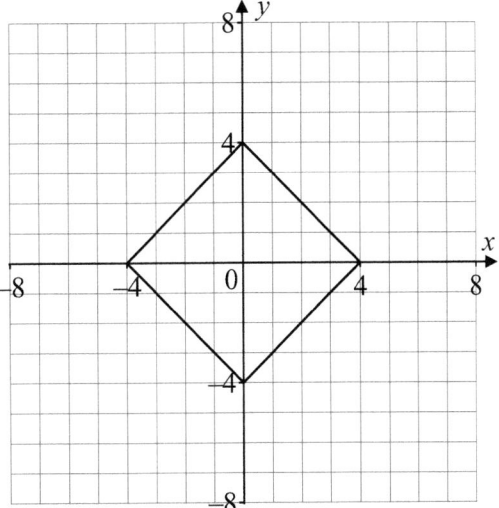

Figure 1

A circle of radius 5 units has its centre at the point (−2, 3).
Find the coordinates of the points of intersection of the circle and the square.
Give your answers correct to 2 decimal places where appropriate.

(7)

8 Given that:

$$f^{-1}(x) = \frac{2x-5}{x}, \quad x \neq 0 \qquad g(x) = \sqrt{2x-k}, \quad x \geq \frac{k}{2}, \text{ where } k \text{ is a positive constant}$$

a) find fg(x), giving your answer in terms of x, and state its domain. **(3)**

b) If gg(10) = 2, find the value of k. **(3)**

9 A curve has equation $3y^2 - 4y = 4 - 2x^3$.

a) Find $\frac{dy}{dx}$ in terms of x and y in its simplest form. (3)

b) Find the exact distance between the two stationary points on the curve. (3)

10 By using the substitution $x = \frac{2}{3}\sin\theta$, find:

$$\int \frac{1}{\sqrt{4-9x^2}}\,dx$$

giving your answer in terms of x.

(5)

11 a) $5\sin\theta + 7\cos\theta$ can be written in the form $R\sin(\theta + \alpha)$ where α is an acute angle.

Find R, giving your answer in surd form, and find α, giving your answer in degrees to the nearest whole number.

(3)

b) Two boats sail from the same harbour on the same bearing.
Boat A sails at 5 mph and boat B sails at 7 mph.

After an hour, Boat A sails due west until it is directly north of where it started, and Boat B sails due south until it is directly east of where it started.
Once both boats are in these positions, the total distance they have sailed is 18 miles.

Find the bearing on which the boats both sailed on the first stage of their journey, giving your answer to the nearest whole degree.

(4)

12 A sum of money is invested in a savings account with a compound interest rate of r%, paid annually. T is the number of years taken for the investment to be worth twice the sum invested.

a) Express T in the form $\log_a(b)$, where b is an integer and a is given in terms of r.
(3)

b) It takes exactly 15 years for an investment to double in an account offering compound interest. Find the interest rate, giving your answer as a percentage correct to 2 decimal places.
(2)

c) Money invested in a simple interest account with annual interest rate p% takes twice as long to double in value as money in a compound interest account with annual interest rate q%.

Show that $q = 100(2^{\frac{p}{50}} - 1)$.
(4)

13. A curve is given by the equation $y = \frac{x^2 \cos x}{3 \sin x}$, $\sin x \neq 0$.

a) Show that $\frac{dy}{dx} = \frac{x(\sin 2x - x)}{3 \sin^2 x}$. (5)

b) The curve has one turning point between $x = 0$ and $x = 1$. The x-coordinate of this turning point is a.

i) Verify that $a = 0.95$ to 2 decimal places. (2)

ii) Determine whether 0.95 is an overestimate or an underestimate for the value of a. (1)

14 Theo models the total number of views (V, in thousands) that each of his videos has t days after he uploads it to his channel using the following equation:

$$\frac{dV}{dt} = \frac{2}{Vt}, \quad t > 0$$

1 day after being uploaded, a particular video has 4000 views.

a) Use the model to estimate how many views the video would get during the 8th day after it was uploaded.

(5)

b) Explain one limitation of this model, and suggest a possible improvement that Theo could make to the model in order to address it.

(2)

15 a) Find $\int \sin^2 5x \, dx$. (4)

b) Hence, find the exact value of $\int_0^{\frac{2\pi}{5}} x \sin^2 5x \, dx$. (5)

END OF QUESTIONS

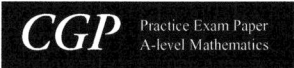

Edexcel A-Level Mathematics

Practice Set 2
Paper 2: Pure Mathematics 2

Time allowed: 2 hours

Centre name					
Centre number					
Candidate number					

Surname	
Other names	
Candidate signature	

In addition to this paper you should have:
- An Edexcel mathematical formula booklet
- A calculator

Instructions to candidates
- Use black ink or ball-point pen.
- A pencil may be used for diagrams, sketches and graphs.
- Write your name and other details in the spaces provided above.
- Answer all questions in the spaces provided.
- Show clearly how you worked out your answers.
- Round answers to 3 significant figures unless otherwise stated.

Information for candidates
- There are 15 questions in this paper.
- There are 100 marks available for this paper.
- The marks available are given in brackets at the end of each question.
- You may get marks for method, even if your answer is incorrect.

Advice to candidates
- Work steadily through the paper and try to answer every question.
- Don't spend too long on one question.
- If you have time at the end, go back and check your answers.

For examiner's use			
Q	Mark	Q	Mark
1		9	
2		10	
3		11	
4		12	
5		13	
6		14	
7		15	
8			
Total			

Exam Set MEP71

© CGP 2018 — copying more than 5% of this paper is not permitted

Answer ALL the questions.

Write your answers in the spaces provided.

1   Point A has position vector $\begin{pmatrix} 4 \\ 1 \\ 5 \end{pmatrix}$ and point B has position vector $\begin{pmatrix} -4 \\ 1 \\ 1 \end{pmatrix}$.

   a) Find the distance between points A and B, giving your answer as a fully simplified surd. (3)

   b) Given that $\overrightarrow{AC} = 3\overrightarrow{AB}$, find the position vector of point C. (2)

2   Given that:
$$64^a \times \left(\frac{1}{16}\right)^b \div \sqrt[c]{32} = 2^d,$$

   express d in terms of a, b and c. (3)

3 A farmer believes that the following formula models the relationship between the amount of money, P, in pounds, she can make from selling her maize crop t days after 1st July:

$$P = -t^2 + 66t + 5069$$

a) With reference to the model, interpret the significance of 5069 in this formula. (1)

b) Use calculus to find the date on which she should sell the crop to make as much money as possible, fully justifying your answer. (4)

c) How much money can she make from the crop if she sells on this date? (1)

d) Describe a limitation of this model and suggest one way in which it could be improved. (2)

4 Solve the equation:

$$4\cos x - 11 = \frac{\sin^2 x - 3}{\cos x}, \quad \cos x \neq 0$$

for $0° \leq x < 360°$. Give your answers correct to 1 decimal place.

(6)

5 a) Sketch the graph of $y = |3x - 1|$ below, clearly stating the coordinates of any points at which the graph touches or intersects the axes.

(3)

b) The equation $|3x - 1| = 2x + 5$ has two solutions.
Use this information to help you solve the inequality $|3x - 1| \leq 2x + 5$.

(3)

6 Sarah runs a weekly maths competition. The number of prizes given out in consecutive competitions forms an arithmetic sequence.

- After 20 competitions, a total of 1390 prizes had been given out.
- After 30 competitions, a total of 3135 prizes had been given out.

How many prizes were given out at the 10th competition?

(5)

7

A curve is defined by the parametric equations:
$$x = 3t + 1 \qquad y = (t+3)^3 - 5.$$

a) Verify that the curve can be written in Cartesian form as $y = \left(\dfrac{x+8}{3}\right)^3 - 5$. **(2)**

From $x = 3t + 1$, $t = \dfrac{x-1}{3}$, so $t + 3 = \dfrac{x-1}{3} + 3 = \dfrac{x-1+9}{3} = \dfrac{x+8}{3}$.

Therefore $y = \left(\dfrac{x+8}{3}\right)^3 - 5$.

b) Caleb wants to evaluate $\displaystyle\int_4^{10} y\, dx$.

He decides to integrate parametrically, and works out that $\displaystyle\int_4^{10} y\, dx = \int_4^{10} ((t+3)^3 - 5)\, dt$.

Identify two errors that Caleb has made. **(2)**

- He has not changed the limits from x-values to t-values. When $x=4$, $t=1$; when $x=10$, $t=3$.
- He has not multiplied by $\dfrac{dx}{dt} = 3$.

c) By correcting Caleb's mistakes, use parametric integration to evaluate $\displaystyle\int_4^{10} y\, dx$. **(3)**

$$\int_4^{10} y\, dx = \int_1^{3} ((t+3)^3 - 5) \cdot 3\, dt = 3\left[\dfrac{(t+3)^4}{4} - 5t\right]_1^3$$

$$= 3\left[\left(\dfrac{6^4}{4} - 15\right) - \left(\dfrac{4^4}{4} - 5\right)\right] = 3[(324 - 15) - (64 - 5)]$$

$$= 3[309 - 59] = 3 \times 250 = 750.$$

8 A scientist is studying a newly-formed ant colony. After observing the colony for several days, the scientist decides that the population can be modelled by the equation $y = at^b$, where y is the population, t is the time that the colony has existed for, measured in days, and a and b are constants.

a) Find a linear equation for $\log_{10} y$ in terms of $\log_{10} t$.

(2)

The scientist plots his observed values of $\log_{10} y$ against $\log_{10} t$, then draws a line of best fit as shown.

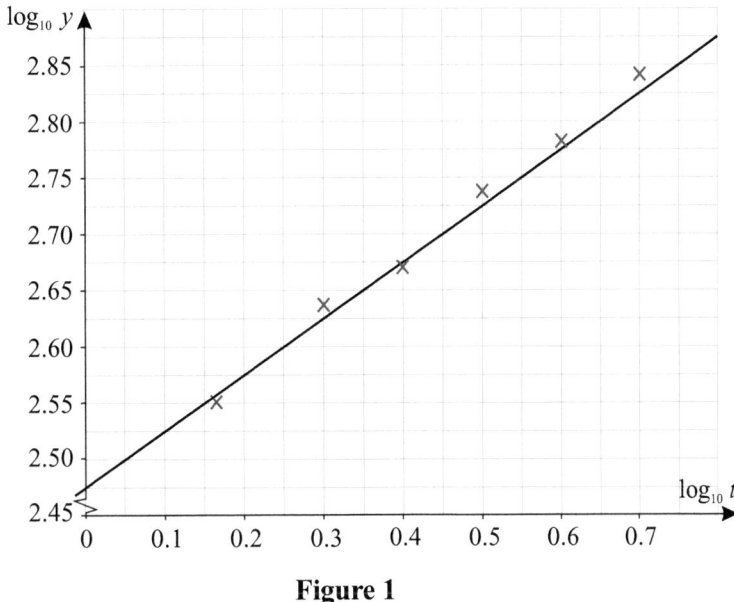

Figure 1

b) By finding the equation of the line of best fit, estimate how many days the colony will have existed for when the population reaches 1000. Give your answer to the nearest whole day.

(4)

c) Give one limitation of this model, explaining your answer in context of the question. **(1)**

9 Figure 2 shows a sketch of the graph of $y = \dfrac{-1}{x}$.

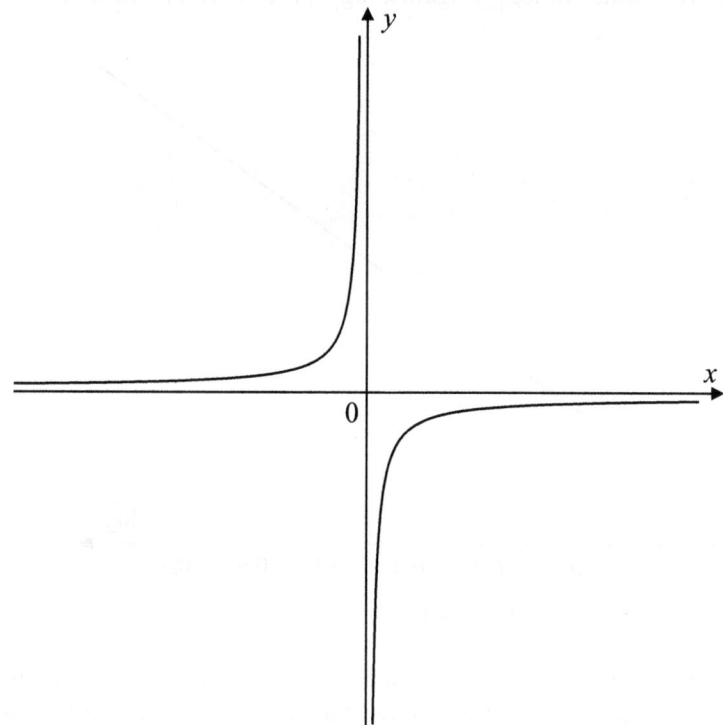

Figure 2

a) Sketch the graph of $y = \dfrac{-1}{x+5} + 2$ on the axes above.
Clearly draw and label any asymptotes with their equations. **(2)**

b) Describe the transformations which would take the graph of $y = \dfrac{-1}{x+5} + 2$ onto the graph of $y = \dfrac{-3}{x+5} + 8$. **(2)**

10 Figure 3 shows part of the curve with equation $y = x^3 - 5x^2 + 2x + 8$ and the line AB. A has coordinates $(0, 0)$. B has coordinates $(1, 6)$ and lies on the curve.

$(-1, 0)$ and C are two points of intersection between the curve and the x-axis.

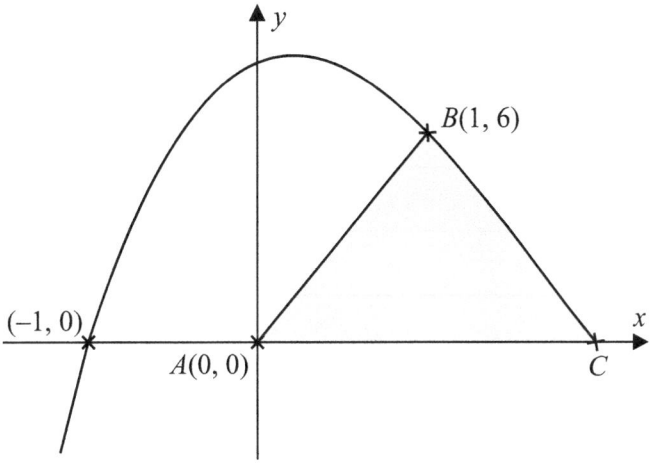

Figure 3

a) Find the coordinates of point C. **(3)**

b) Find the exact area of the shaded region. **(4)**

11 a) Prove, from first principles, that the derivative of x^3 is $3x^2$. (4)

b) The curve $y = \sin^3 x$ has a stationary point at $x = \frac{\pi}{2}$.

By finding $\frac{d^2y}{dx^2}$, determine the nature of the stationary point. (5)

12 Figure 4 shows a circle with centre O and radius r, divided into two segments, S_1 and S_2, by the line AB.

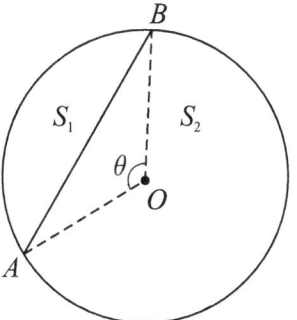

Diagram not drawn accurately.

Figure 4

θ is the obtuse angle AOB, given in radians.

a) Given that the ratio of the area of S_1 to the area of S_2 is $2:7$, show that:

$$\theta - \sin\theta - \frac{4\pi}{9} = 0$$

(4)

b) Using the iterative formula $\theta_{n+1} = \sin\theta_n + \frac{4\pi}{9}$ and $\theta_0 = \frac{\pi}{2}$, find the value of θ correct to 2 significant figures, justifying your answer.

(2)

13 a) Express $f(x) = \dfrac{-x-8}{x^2+6x+8}$ as partial fractions. **(3)**

b) Hence find the binomial expansion of $f(x)$, in ascending powers of x, as far as the term in x^2. Give your answer in its fully simplified form. **(4)**

c) Find the range of values of *x* for which this expansion is valid.
Give your answer in set notation. **(3)**

d) Find the percentage error when your expansion is used to find an estimate
for the value of f(0.1). Give your answer correct to 2 significant figures. **(2)**

14 a) Prove that $\dfrac{1-\tan^2\theta}{1+\tan^2\theta} \equiv \cos 2\theta$. **(4)**

b) Hence find the exact solutions of the equation:

$$\frac{1-\tan^2\left(\beta+\frac{\pi}{2}\right)}{1+\tan^2\left(\beta+\frac{\pi}{2}\right)} - 0.5\sec(2\beta+\pi) = 0, \quad \cos(2\beta+\pi) \neq 0, \text{ for } 0 < \beta < \pi.$$

(5)

Leave blank

15 The number of customers at an ice cream parlour decreases after the end of the summer holidays. The rate of change of the decrease can be modelled by the differential equation $\frac{dC}{dt} = -kCt$, where:

- $k > 0$ is a constant
- C is the number of customers per week
- t is the number of weeks after the end of the holidays

a) At the end of the holidays, the ice cream parlour had 3600 customers per week. Use this information to solve the differential equation, giving your answer as a formula for C in terms of k and t.

(3)

b) The ice cream parlour will close when the number of customers drops below 300 per week. Given that the value of k is 0.2, calculate how many weeks after the end of the holidays the ice cream parlour will close. Give your answer to the nearest whole week.

(3)

END OF QUESTIONS

Edexcel A-Level Mathematics

Practice Set 2
Paper 3: Statistics and Mechanics

Time allowed: 2 hours

Centre name	
Centre number	
Candidate number	

Surname	
Other names	
Candidate signature	

In addition to this paper you should have:
- An Edexcel mathematical formula booklet
- A calculator

Instructions to candidates
- Use black ink or ball-point pen.
- A pencil may be used for diagrams, sketches and graphs.
- Write your name and other details in the spaces provided above.
- Answer all questions in the spaces provided.
- Show clearly how you worked out your answers.
- Round answers to 3 significant figures unless otherwise stated.

Information for candidates
- There are 12 questions in this paper.
- There are 100 marks available for this paper.
- The marks available are given in brackets at the end of each question.
- You may get marks for method, even if your answer is incorrect.

For examiner's use

Q	Mark	Q	Mark
1		7	
2		8	
3		9	
4		10	
5		11	
6		12	
Total			

Advice to candidates
- Work steadily through the paper and try to answer every question.
- Don't spend too long on one question.
- If you have time at the end, go back and check your answers.

Exam Set MEP71

Answer ALL the questions.

Write your answers in the spaces provided.

SECTION A: STATISTICS

1 Tanya is investigating the daily mean windspeeds in Leeming.

The grouped frequency table below summarises the daily mean windspeeds in Leeming from 22nd September to 31st October 1987.

Windspeed (nearest kn)	Frequency
0 – 4	17
5 – 8	14
9 – 12	7
13 – 18	2

a) Tanya says that the class width of the 5 – 8 category is 3 kn.
Explain why she is not correct.
(1)

b) Use the data to estimate the lower quartile (Q_1), upper quartile (Q_3) and interquartile range (IQR) for the daily mean windspeeds over this period.
(4)

c) Tanya considers any value outside the boundaries below to be an outlier:

Lower boundary = $Q_1 - (1.5 \times IQR)$ Upper boundary = $Q_3 + (1.5 \times IQR)$

Given that the lowest value in the data is 1 kn and the highest value is 18 kn, use these boundaries to determine whether either of these values are outliers.

(2)

d) Tanya uses her calculator to estimate the mean of the data.

Explain whether you think the mean or the median would be the better average to use to represent the data.

(1)

e) Rezan is investigating the daily mean windspeeds for Leuchars over the same period. He calculates the mean of his data, and compares it to the mean of Tanya's data.

The mean of Rezan's data is significantly higher than that of Tanya's data. Rezan claims that this suggests that Scotland is generally windier than England.

Give two criticisms of Rezan's conclusion.

(2)

2 A teacher surveys 100 students to find out whether they like watching reality TV, soaps and talent shows. The results are summarised below:

- 13 only like watching reality TV.
- 26 like watching soaps and reality TV.
- 40 like watching soaps.
- 17 do not like watching any of the shows.
- 11 enjoy just reality TV and talent shows.
- 5 enjoy just soaps and talent shows.
- 4 like just soaps and reality TV.

a) Draw a Venn diagram to show the number of elements in R, S and T, where R is {students that like reality TV}, S is {students that like soaps}, and T is {students that like talent shows}.

(4)

b) A student is picked at random from the 100 students surveyed.

  i) What is the probability that this student likes at least one of the three types of shows?

(2)

ii) Given that the student likes soaps, what is the probability that they like talent shows? **(2)**

iii) Given that the student likes talent shows, what is the probability that they like reality TV but not soaps? **(2)**

c) The teacher says, "the probability that a student enjoys talent shows is not affected by whether or not they enjoy soaps."

Do the results of the survey support this claim? Show your working. **(3)**

3 The scatter graph in Figure 1 shows the mean temperature (T °C) and the maximum relative humidity (H%) for 10 days in Heathrow in 1987.

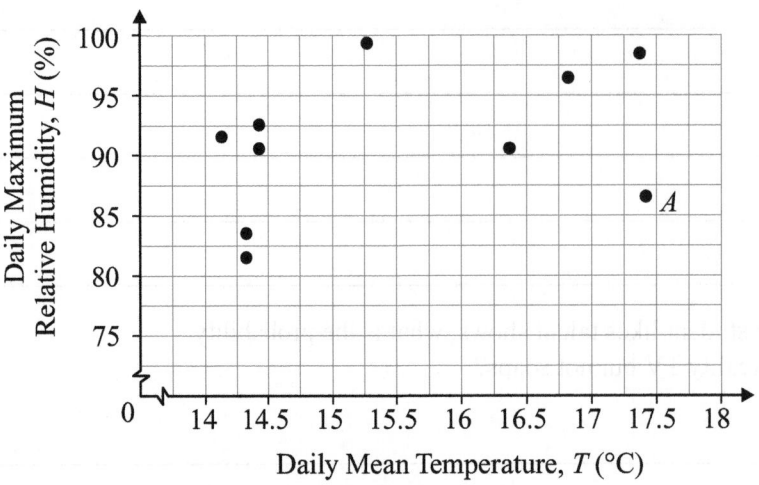

Figure 1

a) Describe the correlation shown in the scatter graph. (1)

b) The product moment correlation coefficient, r, is calculated to be 0.3850, correct to 4 d.p. Stating your hypotheses clearly, perform a hypothesis test at the 10% significance level to test whether the product moment correlation coefficient between daily mean temperature and daily maximum relative humidity in Heathrow in 1987 is non-zero. (3)

c) The point labelled A is removed from the sample. What effect this will have on the value of r? (1)

4 A confectionery company offers a prize to anyone who finds a gold-wrapped sweet in a packet of their boiled sweets. The company say that they have randomly selected 9 out of every 20 packets of sweets, and hidden one gold-wrapped sweet in each.

a) Veronica buys 50 packets of these sweets. G is the number of gold-wrapped sweets in the 50 packets that she has bought. Give one way in which G meets the conditions for following a binomial distribution and state the parameters of the distribution.
(2)

After opening all 50 packets, Veronica finds that only 16 of them contained a gold-wrapped sweet. She claims that the confectionery company has lied, and that they must have put gold-wrapped sweets into fewer packets than they said.

b) Carry out a hypothesis test at the 5% significance level to determine whether Veronica's findings provide statistically significant evidence that her claim is true. State your hypotheses clearly.
(4)

5 It is known that 3% of the UK's population carry a certain disease.
A test for the disease is available which always gives either a positive or a negative result.

Given that the individual carries the disease,
there is a 98% chance that the test will give a positive result.

Given that the individual does not carry the disease,
there is a 95% chance that the test will give a negative result.

a) **i)** Joey takes the test and gets a positive result.
Find the probability that he carries the disease. (4)

ii) State an assumption that you have made. (1)

b) Hiroshi claims that the test will give the correct result (98% + 95%) ÷ 2 = 96.5% of the time.
Is he correct? Show working to support your answer. (2)

c) Joey and Hiroshi disagree about whether the test is effective. By considering your answers to parts a) and b), explain whether you think this is a good test to see if people carry the disease. (1)

6 A factory manufactures pencils and the plastic boxes that they are stored in. The lengths of the pencils made in the factory are normally distributed with mean 18 cm and standard deviation 0.1 cm.

The plastic boxes used to store the pencils are 18.2 cm long. 5 random pencils are put into each box. If any of the pencils do not fit in the box, the box will be damaged.

a) Find the probability that a box will be damaged.
You may assume that the sides of the box have negligible thickness.

(4)

b) The factory produces 1000 boxes of pencils each day. Using a normal approximation, estimate the probability that more than 125 boxes will be damaged in a single day.

(4)

SECTION B: MECHANICS

[Unless otherwise indicated, give your answers to either 2 or 3 significant figures, and take g = 9.8 ms⁻².]

7 A non-uniform plank, *AB*, of length 4 m and weight *W* N is modelled as a rigid rod.
It rests on a support at its midpoint.

A child (modelled as a particle) whose weight is $\frac{5}{8}$ the weight of the plank,
sits at a point 1 m from *B*, as shown in Figure 2.

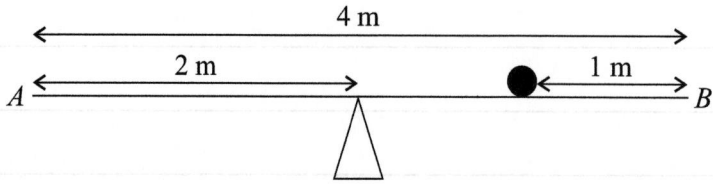

Figure 2

Given that the plank is in equilibrium, find:

a) the magnitude of the reaction force from the support, *R*, in terms of *W*, (2)

b) the distance of the plank's centre of mass from *A*. (3)

8 *[In this question, **i** and **j** are horizontal unit vectors in the East and North directions respectively.]*

A large park area is used for leisure activities. At $t = 0$ seconds, a skateboarder sets off from the origin with constant velocity $(1.5\mathbf{i} - 2\mathbf{j})$ ms^{-1}.

a) Find the skateboarder's position vector, relative to the origin, after 7 seconds. **(2)**

b) After 7 seconds, the skateboarder continues to move at the same velocity and a cyclist sets off from point A with initial velocity $(-4.5\mathbf{i} + 6\mathbf{j})$ ms^{-1} and constant acceleration $(-0.3\mathbf{i} + 0.4\mathbf{j})$ ms^{-2}. The two continue to move in this way and collide 6 seconds later.

Find the position vector, relative to the origin, of:

i) the point where they collide, **(2)**

ii) point A. **(3)**

9 *[In this question, give your answers to 2 significant figures.]*

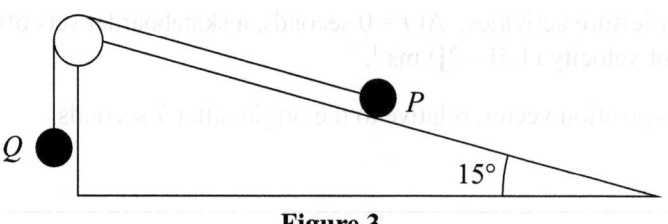

Figure 3

Figure 3 shows two particles, P and Q. P has a mass of 2 kg and Q has a mass of 4 kg.

The particles are connected by a light, inextensible string which passes over a smooth pulley at the top of the plane.

The section of the string which connects P to the pulley is parallel to the line of greatest slope of the plane. The plane is smooth and inclined at 15° to the horizontal.

Particle P is initially held at rest while particle Q hangs freely. Particle P is then released.

a) Find the magnitude of the acceleration of the system. **(5)**

b) The tension in the string exerts a force, F, on the pulley.
Find the magnitude of the vertical component of F. **(3)**

10 A particle moves with velocity $\mathbf{v} = \begin{pmatrix} 5t - 6t^2 \\ 6 - t^3 \end{pmatrix}$ ms^{-1}, where t is the time, given in seconds.

When $t = 0$, the position vector of the particle relative to the origin, is $\begin{pmatrix} 6 \\ 2 \end{pmatrix}$ m.

a) Find the position vector of the particle after 3 seconds. (4)

b) Find the magnitude of the acceleration of the particle when $t = 1$ second. (4)

c) Given that the particle has a mass of 5 kg, find the value of k when the resultant force on the particle is equal to $\begin{pmatrix} -20 \\ k \end{pmatrix}$ N. (3)

11 A bullet is fired upwards, at an angle of $\theta°$ with the horizontal, from a height of 3.5 m above the horizontal ground. The initial speed of the bullet is 203 ms^{-1} and it hits the ground after 5 seconds.

    a) By modelling the bullet as a particle moving freely under gravity, find θ. (3)

    b) Find the angle between the horizontal and the direction of the bullet's motion as it hits the ground. (4)

    c) After how many seconds was the bullet travelling at its minimum speed? (2)

12 A particle with mass 2 kg is held at rest on a rough sloping plane.
The angle between the line of greatest slope of the plane and the horizontal is 20°.
When released, the particle moves down the plane with an acceleration of $0.15g$ ms^{-2}.

a) Find the coefficient of friction, μ, between the particle and the plane, giving your answer correct to 2 d.p.

(6)

b) The plane is then adjusted so that the angle between its line of greatest slope and the horizontal is $\theta°$. When the particle is released, it remains in equilibrium. Find the range of possible values of θ.

(4)

END OF QUESTIONS

CGP

A-Level

Mathematics

Exam Board: Edexcel

Practice Exam Papers
Instructions & Answer Booklet

Exam Set MEP71

CGP Practice Papers are integral to A-Level Maths exam prep!

Get ready for some serious calculus, because this brilliant pack from CGP contains two full sets of Practice Papers for Edexcel A-Level Maths.

They're carefully crafted to be just like the real exams — so you'll know exactly what to expect on the day. We've also included fully worked answers and clear mark schemes for each question. What more could you ask?

(Actually, you could ask for a formula booklet. But you don't have to, because we've given you one of those too!)

CGP — still the best! ☺

Our sole aim here at CGP is to produce the highest quality books — carefully written, immaculately presented and dangerously close to being funny*.

Then we work our socks off to get them out to you
— at the cheapest possible prices.

*Admittedly these practice papers aren't very funny, since we were too busy concentrating on making them as serious as the real exams. But normally we'd include more jokes, honestly.

Marking Your Papers

- Do a complete exam (Paper 1, Paper 2 and Paper 3).
- Use the answers and mark scheme in this booklet to mark each exam paper.
- Write down your mark for each paper in the table below — each paper is worth a maximum of 100 marks.
- Find your total for the whole exam (out of a maximum of 300 marks) by adding up your marks from all three papers.
- Follow the instructions below to estimate your grade.

	Paper 1	Paper 2	Paper 3	Total	Grade
SET 1					
SET 2					

Estimating Your Grade

- If you want to get a **rough idea** of the grade you're working at, we suggest you compare the **total mark** you got in **each set** to the latest set of grade boundaries.
- Grade boundaries are set for each individual exam, so they're likely to **change** from year to year. You can find the latest set of grade boundaries by going to **www.cgpbooks.co.uk/alevelgradeboundaries**
- Jot down the marks required for each grade in the table below so you don't have to refer back to the website. Use these marks to **estimate your grade**. If you're borderline, don't push yourself up a grade — the real examiners won't.

Total mark required for each grade					
Grade	A*	A	B	C	D	E
Total mark out of 300						

- Remember, this will only be a **rough guide**, and grade boundaries will be different for different exams, but it should help you to see how you're getting on.

Published by CGP

Editors: Sammy El-Bahrawy and Shaun Harrogate.

Contributors: Paul Garrett and Charlotte Young.

With thanks to Allan Graham and Dawn Wright *for the proofreading.*

Printed by Elanders Ltd, Newcastle upon Tyne.
ISBN: 978 1 78908 063 6
Text, design, layout and original illustrations
© Coordination Group Publications Ltd. (CGP) 2018
All rights reserved.

Set 1 Paper 3, Pages 2-5 and Set 2 Paper 3, Pages 2, 3 & 6 © Crown Copyright, the Met Office.
Contains public sector information licensed under the Open Government Licence v3.0 ~
http://www.nationalarchives.gov.uk/doc/open-government-licence/version/3/

Photocopying more than 5% of a paper is not permitted, even if you have a CLA licence.
Extra copies are available from CGP with next day delivery • 0800 1712 712 • www.cgpbooks.co.uk

Answers

Set 1 Paper 1 — Pure Mathematics 1

1. a) f'(x) = $3x^2 - 8x - 3$
   [2 marks available — 1 mark for differentiating to get a quadratic, 1 mark for the correct answer]

   b) f(x) is decreasing when f'(x) < 0:
   f'(x) = 0 $\Rightarrow$ $3x^2 - 8x - 3 = 0$ $\Rightarrow$ $(3x + 1)(x - 3) = 0$ *[1 mark]*
   $\Rightarrow x = -\frac{1}{3}$ and 3 *[1 mark]* f'(x) is u-shaped so is negative when $-\frac{1}{3} < x < 3$. So in the required set notation, this is $\{x : x > -\frac{1}{3}\} \cap \{x : x < 3\}$ *[1 mark]*.
   [3 marks available in total — as above]

2.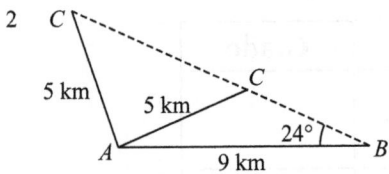

   $\frac{\sin ACB}{9} = \frac{\sin 24°}{5}$ $\Rightarrow$ $\sin ACB = \frac{9\sin 24°}{5} = 0.7321...$
   $\Rightarrow$ Angle ACB = $\sin^{-1}(0.7321...) = 47.06...°$ or $180° - 47.06...°$
   $\Rightarrow$ Angle ACB = 47.1° (1 d.p.) or Angle ACB = 132.9° (1 d.p.)
   [3 marks available — 1 mark for using the sine rule correctly, 1 mark for 47.1°, 1 mark for 132.9°]

3. a) $(\sqrt{2}x^{\frac{1}{3}})^6 = \sqrt{2}^6 x^{\frac{1}{3}\times 6} = 8x^2$
   [2 marks available — 1 mark for an answer in the form kx^2 or $8x^k$, 1 mark for the correct answer]

   b) Substitute $y = 1 - 3x$ into $8x^2 + 6x - 18 = y^2 - y$ to give
   $8x^2 + 6x - 18 = (1 - 3x)^2 - (1 - 3x)$ *[1 mark]*
   $\Rightarrow 8x^2 + 6x - 18 = 1 - 6x + 9x^2 - 1 + 3x$ *[1 mark]*
   $\Rightarrow x^2 - 9x + 18 = 0$
   $\Rightarrow (x - 3)(x - 6) = 0$ *[1 mark]*
   $\Rightarrow x = 3$ and $x = 6$ *[1 mark]*
   [4 marks available in total — as above]

4. a) E.g. in a regular 3-sided shape (equilateral triangle) the exterior angles are 360° ÷ 3 = 120° which is not acute *[1 mark]*.
   You could also have chosen a regular 4-sided shape (square) because the exterior angles are 360° ÷ 4 = 90° — not acute.

   b) Choose any value of n such that $-1 < n < 0$, e.g. $n = -0.2$
   $\Rightarrow \frac{n}{n+1} = \frac{-0.2}{-0.2+1} = \frac{-0.2}{0.8} = -0.25$ which is less than 0.
   [1 mark]

   c) E.g. $n = 23 \Rightarrow n + 2 = 25$, $n + 4 = 27$ and 25 and 27 are not prime *[1 mark]*.
   You could also have chosen n = 31 and n = 47.

5. $E \propto x^2 \Rightarrow E = kx^2 \Rightarrow 7\sqrt{2} - 9 = k(\sqrt{2} - 1)^2$ *[1 mark]*
   $(\sqrt{2} - 1)^2 = 2 - 2\sqrt{2} + 1 = 3 - 2\sqrt{2}$ *[1 mark]*
   $k = \frac{7\sqrt{2} - 9}{3 - 2\sqrt{2}} \times \frac{3 + 2\sqrt{2}}{3 + 2\sqrt{2}}$ *[1 mark]*
   $= \frac{21\sqrt{2} - 27 + 28 - 18\sqrt{2}}{9 - 8} = 1 + 3\sqrt{2}$ *[1 mark]*
   When $x = \sqrt{2}$ m,
   $E = (1 + 3\sqrt{2})(\sqrt{2})^2 = 2(1 + 3\sqrt{2}) = 2 + 6\sqrt{2}$ joules *[1 mark]*
   [5 marks available in total — as above]

6. a) i) $x^2 - 8x + y^2 + 4y - 29 = 0$
      $\Rightarrow (x - 4)^2 - 16 + (y + 2)^2 - 4 - 29 = 0$ *[1 mark]*
      $\Rightarrow (x - 4)^2 + (y + 2)^2 = 49$ $\Rightarrow$ centre X is (4, −2) *[1 mark]*
      [2 marks available in total — as above]

   ii) Radius = $\sqrt{49} = 7$ *[1 mark]*
       You could also have put the circle equation into the form $x^2 + 2gx + y^2 + 2fy + c = 0$ and then the centre is at $(-g, -f)$ and the radius is $\sqrt{g^2 + f^2 - c}$.

   b) P (−16, 13), X (4, −2) and Y form a right-angled triangle, because a tangent (PY) meets a radius (XY) at 90°.
   Length of PX = $\sqrt{(-16-4)^2 + (13-(-2))^2}$
   $= \sqrt{(-20)^2 + 15^2} = \sqrt{625} = 25$ *[1 mark]*
   Length of XY = radius = 7
   $PY^2 = PX^2 - XY^2 \Rightarrow PY^2 = 625 - 7^2$ *[1 mark]*
   $\Rightarrow PY^2 = 576 \Rightarrow PY = \sqrt{576} = 24$ *[1 mark]*
   [3 marks available in total — as above]

7. a) $\frac{4x^2 + 8x + 3}{4x^2 - 9} = \frac{(2x+3)(2x+1)}{(2x+3)(2x-3)} = \frac{2x+1}{2x-3}$
   [2 marks available — 1 mark for factorising the numerator or denominator correctly, 1 mark for the correct answer]

   b) $\log_3(4x^2 + 8x + 3) - \log_3(4x^2 - 9) = 2$
   $\Rightarrow \log_3\left(\frac{4x^2+8x+3}{4x^2-9}\right) = 2$ *[1 mark]*
   $\Rightarrow \log_3\left(\frac{2x+1}{2x-3}\right) = 2 \Rightarrow \frac{2x+1}{2x-3} = 3^2$ *[1 mark]*
   $\Rightarrow 2x + 1 = 9(2x - 3) \Rightarrow 2x + 1 = 18x - 27$ *[1 mark]*
   $\Rightarrow 16x = 28 \Rightarrow x = \frac{7}{4}$ or 1.75 *[1 mark]*
   [4 marks available in total — as above]

8. a) Point A is a (stationary) point of inflection *[1 mark]* as e.g. the graph goes from convex to concave *[1 mark]*.
   [2 marks available in total — as above]

   b) f(x) doesn't have an inverse (unless the domain is restricted) because f(x) is many-to-one, not one-to-one. *[1 mark]*
   Alternatively, you could have said that $f^{-1}(x)$ cannot be one-to-one, or that $f^{-1}(x)$ would be one-to-many.

   c) i) f(x − 1) is a translation of $\binom{1}{0}$
      $\Rightarrow$ B has coordinates (p + 1, q) *[1 mark]*

      ii) 3f(2x) is a stretch parallel to the x-axis of scale factor $\frac{1}{2}$ then a stretch parallel to the y-axis of scale factor 3
      $\Rightarrow$ B has coordinates $\left(\frac{p}{2}, 3q\right)$ *[1 mark]*

   d) i) f'(x) = $12x^3 - 6x^2$ *[1 mark]*
      The turning points are found at f'(x) = 0
      $\Rightarrow 6x^2(2x - 1) = 0$ *[1 mark]* $\Rightarrow x = 0$ and $\frac{1}{2}$
      From the graph, p is positive, so $p = \frac{1}{2}$ *[1 mark]*
      and $q = 3\left(\frac{1}{2}\right)^4 - 2\left(\frac{1}{2}\right)^3 - 2 = -\frac{33}{16}$ or −2.0625 *[1 mark]*
      [4 marks available in total — as above]

      ii) f"(x) = $36x^2 - 12x$ *[1 mark]*
      When $x = \frac{1}{2}$, f"(x) = $36\left(\frac{1}{2}\right)^2 - 12\left(\frac{1}{2}\right) = 3$, which is greater than 0 so B is a minimum *[1 mark]*.
      [2 marks available in total — as above]

   e) $y = |f(-x)|$ is a reflection of $y = f(x)$ in the y-axis, and a reflection in the x-axis of the part of the graph where $y < 0$.

   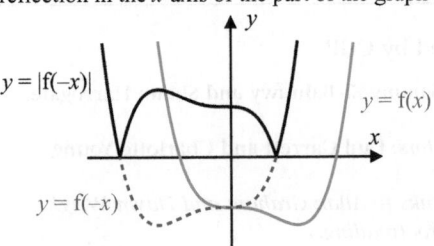

   [2 marks available — 1 mark for each correct transformation]

9. Let the radius of the cylinder be r.
   Then the area of triangle ABC = $\frac{1}{2}r^2 \sin 1.02 = 85.9$ *[1 mark]*
   $\Rightarrow r = \sqrt{\frac{2 \times 85.9}{\sin 1.02}} = 14.199...$ cm *[1 mark]*
   Arc length = $r\theta$ = 14.199... × 1.02 = 14.483... cm *[1 mark]*

Area of wax = arc length × h = 14.483... × 4
= 57.932... cm² = 57.9 cm² (3 s.f.) *[1 mark]*
[4 marks available in total — as above]

10 a) E.g. f(x) might have an asymptote / might not be continuous between a and b. *[1 mark]*
   E.g. if f(x) = $\frac{1}{x}$, f(–1) = –1 < 0 and f(1) = 1 > 0 but there is no root of f(x) = 0 between –1 and 1.

   b) i) The student prematurely rounded the answer for x_1 when finding x_2 *[1 mark]* and they have concluded that the root is at 3.613 (3 d.p.) before checking that this agrees with the next value of x_n. *[1 mark]*
   [2 marks available in total — as above]

   ii) x_2 = 3.6176... x_3 = 3.5919... x_4 = 3.5917...
   So α = 3.592 (3 d.p.)
   You can stop here because both 3.5919... and 3.5917... round to 3.592 (3 d.p.).
   [2 marks available — 1 mark for iterating until two values round to the same number to 3 d.p., 1 mark for the correct final answer]

11 a) This is a geometric sequence with r = 1.05
   The nth term = ar^{n-1} = 250 000 × 1.05^{n-1} > 500 000
   $\Rightarrow 1.05^{n-1} > 2 \Rightarrow (n-1)\ln 1.05 > \ln 2$
   $\Rightarrow n - 1 > 14.2... \Rightarrow n > 15.2... \Rightarrow n = 16$
   n = 1 in 2021, so the expected profit will be over £500 000 in 2036.
   [4 marks available — 1 mark for the correct values of r and a, 1 mark for setting up an inequality correctly, 1 mark for solving by taking logs of both sides, 1 mark for the correct year]

   b) a = 250 000, r = 1.05, n = 10 so since $S_n = \frac{a(1-r^n)}{1-r}$ then
   $S_{10} = \frac{250\,000(1-1.05^{10})}{1-1.05}$ *[1 mark]*
   = 3 144 473.13... = £3 144 500 (nearest £100) *[1 mark]*
   [2 marks available in total — as above]

12 a) $\frac{18x-7}{(2-3x)(1+x)} \equiv \frac{A}{2-3x} + \frac{B}{1+x}$
   $\Rightarrow 18x - 7 \equiv A(1+x) + B(2-3x)$.
   Using $x = \frac{2}{3} \Rightarrow 12 - 7 = \frac{5}{3}A \Rightarrow A = 3$
   and $x = -1 \Rightarrow -18 - 7 = 5B \Rightarrow B = -5$
   So $\frac{18x-7}{(2-3x)(1+x)} \equiv \frac{3}{2-3x} - \frac{5}{1+x}$
   [3 marks available — 1 mark for splitting the expression into two fractions with the correct denominators, 1 mark for a correct method to find A or B, 1 mark for the correct answer]

   b) $\frac{18x-7}{(2-3x)(1+x)} = \frac{3}{2-3x} - \frac{5}{1+x}$
   $= 3(2-3x)^{-1} - 5(1+x)^{-1} = \frac{3}{2}\left(1-\frac{3}{2}x\right)^{-1} - 5(1+x)^{-1}$ *[1 mark]*
   $\left(1-\frac{3}{2}x\right)^{-1} = 1 + (-1)\left(-\frac{3}{2}x\right) + \frac{-1 \times -2}{2!}\left(-\frac{3}{2}x\right)^2 + ...$ *[1 mark]*
   $= 1 + \frac{3}{2}x + \frac{9}{4}x^2 + ...$ *[1 mark]*
   So $\frac{3}{2}\left(1-\frac{3}{2}x\right)^{-1} - 5(1+x)^{-1}$
   $= \frac{3}{2}\left(1 + \frac{3}{2}x + \frac{9}{4}x^2 + ...\right) - 5(1-x+x^2+...)$ *[1 mark]*
   $= \frac{3}{2} + \frac{9}{4}x + \frac{27}{8}x^2 - 5 + 5x - 5x^2 + ...$
   $= -\frac{7}{2} + \frac{29}{4}x - \frac{13}{8}x^2 + ...$ *[1 mark]*
   [5 marks available in total — as above]

13 Using $\sin 2\theta \equiv 2\sin\theta\cos\theta$ and $\cos 2\theta \equiv 2\cos^2\theta - 1$
   $\frac{\sin 2\theta + \sin\theta}{\cos 2\theta + 1 + \cos\theta} = \frac{2\sin\theta\cos\theta + \sin\theta}{(2\cos^2\theta - 1) + 1 + \cos\theta}$
   $= \frac{\sin\theta(2\cos\theta + 1)}{\cos\theta(2\cos\theta + 1)} = \frac{\sin\theta}{\cos\theta} = \tan\theta$
   [4 marks available — 1 mark for using the identity for sin 2θ, 1 mark for using the identity for cos 2θ, 1 mark for factorising the expression, 1 mark for the correct identity for tan θ]

14 a) Let $u = 1 + 3e^{-x} \Rightarrow \frac{du}{dx} = -3e^{-x} \Rightarrow du = -3e^{-x}dx$
   So $\int \frac{75e^{-x}}{(1+3e^{-x})^2}dx = \int \frac{-25(-3e^{-x})}{(1+3e^{-x})^2}dx$
   $= \int \frac{-25}{u^2}du = \int -25u^{-2}du = -25(-u^{-1}) + C = \frac{25}{u} + C$
   so $\int \frac{75e^{-x}}{(1+3e^{-x})^2}dx = \frac{25}{1+3e^{-x}} + C$ (i.e. $k = 25$)
   [5 marks available — 1 mark for any suitable substitution, 1 mark for differentiating the substitution with respect to x, 1 mark for forming a new integral in terms of u, 1 mark for a correct method to solve the integral in terms of u, 1 mark for the correct answer in terms of x]

   b) $\frac{dP}{dt} = \frac{75e^{-t}}{2P(1+3e^{-t})^2} \Rightarrow 2P\,dP = \frac{75e^{-t}}{(1+3e^{-t})^2}dt$
   $\Rightarrow \int 2P\,dP = \int \frac{75e^{-t}}{(1+3e^{-t})^2}dt$ *[1 mark]*
   $\Rightarrow P^2 = \frac{25}{1+3e^{-t}} + C$ *[1 mark]*
   When $t = 0$, $P = 5.5 \Rightarrow 5.5^2 = \frac{25}{1+3} + C$ *[1 mark]*
   $\Rightarrow C = 30.25 - 6.25$
   $\Rightarrow C = 24$ so $P^2 = \frac{25}{1+3e^{-t}} + 24$ *[1 mark]*
   [4 marks available in total — as above]

   c) As $t \to \infty$, $e^{-t} = \frac{1}{e^t} \to 0 \Rightarrow P^2 \to \frac{25}{1+3(0)} + 24$ *[1 mark]*
   $\Rightarrow P^2 \to 49 \Rightarrow P \to 7$
   So the limit of the size of the population is 7000 *[1 mark]*.
   [2 marks available in total — as above]

15 Let the side length of the cube be x cm.
   $V = x^3 \Rightarrow \frac{dV}{dx} = 3x^2$ *[1 mark]*
   $A = 6x^2 \Rightarrow \frac{dA}{dx} = 12x$ *[1 mark]*
   Using $\frac{dV}{dt} = \frac{dV}{dx} \times \frac{dx}{dt}$ with $\frac{dV}{dt} = 1.5$ and $\frac{dV}{dx} = 3x^2$
   gives $1.5 = 3x^2 \times \frac{dx}{dt} \Rightarrow \frac{dx}{dt} = \frac{1}{2x^2}$ *[1 mark]*
   Using $\frac{dA}{dt} = \frac{dA}{dx} \times \frac{dx}{dt}$ with $\frac{dx}{dt} = \frac{1}{2x^2}$ and $\frac{dA}{dx} = 12x$
   gives $\frac{dA}{dt} = 12x \times \frac{1}{2x^2} = \frac{6}{x}$ *[1 mark]*
   The cube will be damaged if $\frac{dA}{dt} > 1 \Rightarrow \frac{6}{x} > 1 \Rightarrow x < 6$.
   So 6 cm is the minimum side length that a cube must have so that it is not damaged when heated. *[1 mark]*.
   [5 marks available in total — as above]

16 a) Differentiating the implicit equation for C with respect to x:
   $3x^2 \to 6x$, $2y^2 \to 4y\frac{dy}{dx}$ and $18 \to 0$ *[1 mark]*
   $-2xy \to -2x\frac{dy}{dx} - 2y$ *[1 mark]*
   So $3x^2 - 2xy + 2y^2 = 18 \Rightarrow 6x - 2x\frac{dy}{dx} - 2y + 4y\frac{dy}{dx} = 0$
   $\Rightarrow (4y - 2x)\frac{dy}{dx} = 2y - 6x$
   $\Rightarrow \frac{dy}{dx} = \frac{y-3x}{2y-x} = \left(\frac{3x-y}{x-2y}\right)$ *[1 mark]*
   When $x = 2$, $3(2)^2 - 2(2)y + 2y^2 = 18 \Rightarrow 12 - 4y + 2y^2 = 18$
   $\Rightarrow y^2 - 2y - 3 = 0$ *[1 mark]* $\Rightarrow (y+1)(y-3) = 0$
   $\Rightarrow y = -1$ and $y = 3$.
   From the graph, the y-coordinate is positive at $x = 2$,
   so line L meets curve C at (2, 3) *[1 mark]*.
   [5 marks available in total — as above]

   b) Gradient of line $L = \frac{dy}{dx}$ at (2, 3) = $\frac{3-3(2)}{2(3)-2} = -\frac{3}{4}$ *[1 mark]*
   so L has equation $y - 3 = -\frac{3}{4}(x - 2)$ *[1 mark]*
   $\Rightarrow y = -\frac{3}{4}x + \frac{9}{2} \Rightarrow 3x + 4y = 18$
   Line L crosses x-axis at $y = 0$
   $\Rightarrow 3x + 4(0) = 18 \Rightarrow 3x = 18 \Rightarrow x = 6$ *[1 mark]*
   Curve C crosses x-axis at $y = 0$
   $\Rightarrow 3x^2 = 18 \Rightarrow x^2 = 6 \Rightarrow x = \pm\sqrt{6} \Rightarrow x = -\sqrt{6}$ *[1 mark]*
   (using the graph, take the negative solution)
   So $w = 6 - (-\sqrt{6}) = 8.449... = 8.45$ (3 s.f.) *[1 mark]*
   [5 marks available in total — as above]

Set 1 Paper 2 — Pure Mathematics 2

1. a) $2^5 = 32$ *[1 mark]*
   b) $2^4 \times {}^5C_1 \times p = 240$ *[1 mark]* $\Rightarrow 80p = 240 \Rightarrow p = 3$ *[1 mark]*
   [2 marks available in total — as above]
   c) $q = 2^3 \times {}^5C_2 \times p^2$ *[1 mark]* $= 8 \times 10 \times 3^2 = 720$ *[1 mark]*
   [2 marks available in total — as above]

2. a)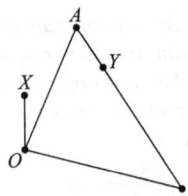
   To get from X to Y, you can go from X to O to A to Y,
   i.e. $\overrightarrow{XY} = \overrightarrow{XO} + \overrightarrow{OA} + \overrightarrow{AY}$ *[1 mark]*
   Y is 3 times further away from B than A, so $\overrightarrow{AY} = \frac{1}{4}\overrightarrow{AB}$
   $\Rightarrow \overrightarrow{AY} = \frac{1}{4}(\overrightarrow{OB} - \overrightarrow{OA}) = \frac{1}{4}((12\mathbf{i} - 3\mathbf{j}) - (4\mathbf{i} + 9\mathbf{j}))$
   $= \frac{1}{4}(8\mathbf{i} - 12\mathbf{j}) = 2\mathbf{i} - 3\mathbf{j}$ *[1 mark]*
   So $\overrightarrow{XY} = \overrightarrow{XO} + \overrightarrow{OA} + \overrightarrow{AY}$
   $= (-4\mathbf{j}) + (4\mathbf{i} + 9\mathbf{j}) + (2\mathbf{i} - 3\mathbf{j})$
   $= (4+2)\mathbf{i} + (-4+9-3)\mathbf{j} = (6\mathbf{i} + 2\mathbf{j})$ km *[1 mark]*
   [3 marks available in total — as above]
   There are lots of ways to do this question, e.g. you could use
   $\overrightarrow{XY} = \overrightarrow{XB} + \frac{3}{4}\overrightarrow{BA}$ or start by finding $\overrightarrow{OY}$. Award the marks for the correct answer with stages of working clearly explained.
   b) Distance $XY = \sqrt{6^2 + 2^2}$ *[1 mark]*
   $= \sqrt{36 + 4} = \sqrt{40} = \sqrt{4}\sqrt{10} = 2\sqrt{10}$ km *[1 mark]*
   [2 marks available in total — as above]

3. a) $t \geq 0$ since time is measured from today, so it cannot have a negative value *[1 mark]*. $t = 9$ is the solution to $S = 0$, so when $t > 9$, S is negative, but you can't have a negative number of subscribers *[1 mark]*.
   [2 marks available in total — as above]
   b) $\frac{dS}{dt} = 7 - 2t$ *[1 mark]* and after 6 months $t = 0.5$
   $\Rightarrow \frac{dS}{dt} = 7 - 2(0.5) = 6$ so the rate of change is 6 million subscribers per year. *[1 mark]*
   [2 marks available in total — as above]
   c) $S = 18 + 7t - t^2 = -(t^2 - 7t - 18)$
   $= -\left(\left(t - \frac{7}{2}\right)^2 - \frac{49}{4} - 18\right)$ *[1 mark]* $= -\left(\left(t - \frac{7}{2}\right)^2 - \frac{121}{4}\right)$
   $= \frac{121}{4} - \left(t - \frac{7}{2}\right)^2$ or $30.25 - (t - 3.5)^2$ *[1 mark]*
   [2 marks available in total — as above]
   d) When $t = 3.5$, $S = 30.25$, i.e. 30 250 000 subscribers *[1 mark]*

4. a) $f(-2) = 0 \Rightarrow x + 2$ is a factor.
   Take out a factor of $(x + 2)$ using e.g. algebraic long division:
   ```
              x² - x - 6
   x + 2 ) x³ + x² - 8x - 12
           x³ + 2x²
           ────────
               -x² - 8x
               -x² - 2x
               ────────
                    -6x - 12
                    -6x - 12
                    ────────
                         0
   ```
   So $f(x) = (x + 2)(x^2 - x - 6)$
   $\Rightarrow f(x) = (x + 2)(x + 2)(x - 3)$ or $(x + 2)^2(x - 3)$
   [4 marks available — 1 mark for using the factor theorem, 1 mark for a correct method to take out a factor of $x + 2$, 1 mark for correctly factorising out $x + 2$, 1 mark for the correct answer]

   b)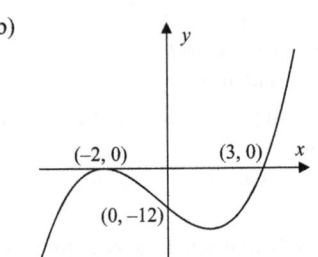
   [3 marks available — 1 mark for the correct shape of a cubic graph with a positive x^3 coefficient, 1 mark for the y-axis intercept of $(0, -12)$, 1 mark for the correct x-axis intercepts]

   c) $g(z) = f(e^{2z})$ and since $e^{2z} > 0$ the only solution is at
   $e^{2z} = 3$ *[1 mark]* $\Rightarrow 2z = \ln 3 \Rightarrow z = \frac{1}{2}\ln 3$ or $\ln\sqrt{3}$ *[1 mark]*
   [2 marks available in total — as above]

5. a) Discriminant $= b^2 - 4ac = k^2 - 4(2)(-2) = k^2 + 16$ *[1 mark]*, which is always positive as $k^2 \geq 0$, so the equation has real solutions for all values of k *[1 mark]*.
   [2 marks available in total — as above]
   b) $k\tan\theta - 2\cos\theta = 0 \Rightarrow k\frac{\sin\theta}{\cos\theta} - 2\cos\theta = 0$ *[1 mark]*
   $\Rightarrow k\sin\theta - 2\cos^2\theta = 0 \Rightarrow k\sin\theta - 2(1 - \sin^2\theta) = 0$ *[1 mark]*
   $\Rightarrow k\sin\theta - 2 + 2\sin^2\theta = 0$
   $\Rightarrow 2\sin^2\theta + k\sin\theta - 2 = 0$ as required *[1 mark]*
   [3 marks available in total — as above]
   c) For $3\tan\theta - 2\cos\theta = 0$, $k = 3$ so
   $2\sin^2\theta + 3\sin\theta - 2 = 0$ *[1 mark]*
   $\Rightarrow (2\sin\theta - 1)(\sin\theta + 2) = 0 \Rightarrow \sin\theta = \frac{1}{2}$ *[1 mark]*
   ($\sin\theta$ cannot equal -2 as $-1 \leq \sin\theta \leq 1$)
   $\Rightarrow \theta = 30°$ and $150°$ *[1 mark]*
   [3 marks available in total — as above]

6. a) $V \propto \frac{1}{P} \Rightarrow V = \frac{k}{P}$
   $V = \pi r^2 h$ so: $\pi r^2 h = \frac{k}{P} \Rightarrow Pr^2 = \frac{k}{\pi h}$ *[1 mark]*
   As h is fixed, $\frac{k}{\pi h}$ is constant.
   Let P_o and r_o be the original pressure and radius.
   Let P_n and r_n be the new pressure and radius.
   $P_o r_o^2 = \frac{k}{\pi h} = P_n r_n^2$ *[1 mark]*
   $r_o = 2$ m and $P_n = (1 - 0.84)P_o = 0.16P_o$ *[1 mark]*
   So: $P_o(2^2) = (0.16P_o)r_n^2 \Rightarrow 4 = 0.16r_n^2$
   $\Rightarrow r_n^2 = 25 \Rightarrow r_n = 5$ m *[1 mark]*
   [4 marks available in total — as above]
   b) Let P_o and r_o be the original pressure and radius respectively, and P_n and r_n be the new pressure and radius. Then $r_n = 2r_o$.
   So: $P_o r_o^2 = P_n r_n^2 \Rightarrow P_o r_o^2 = P_n(2r_o)^2$ *[1 mark]*
   $\Rightarrow P_o r_o^2 = 4P_n r_o^2 \Rightarrow P_o = 4P_n \Rightarrow P_n = \frac{1}{4}P_o$
   So P_n is $\frac{1}{4} = 25\%$ of P_o, which is a 75% decrease. *[1 mark]*
   [2 marks available in total — as above]

7. a) $u_3 = pu_2 - 3$ so $33 = 45p - 3 \Rightarrow 45p = 36 \Rightarrow p = \frac{4}{5}$
   $u_2 = \frac{4}{5}u_1 - 3 \Rightarrow \frac{4}{5}u_1 = 45 + 3 = 48 \Rightarrow u_1 = 60$
   $u_4 = \frac{4}{5}u_3 - 3 = \frac{4}{5}(33) - 3 = \frac{117}{5}$ or $23\frac{2}{5}$, so required sum is
   $60 + 45 + 33 + \frac{117}{5} = \frac{807}{5}$ or $161\frac{2}{5}$
   [3 marks available — 1 mark for finding the value of p, 1 mark for finding the value of u_1 or u_4, 1 mark for the correct answer]
   b) As n tends to infinity you can assume that $u_n = u_{n+1} = L$
   $L = \frac{4}{5}L - 3$ *[1 mark]* $\Rightarrow \frac{1}{5}L = -3 \Rightarrow L = -15$ *[1 mark]*
   [2 marks available in total — as above]

8 a) $n = \dfrac{1 - 0.4}{3} = \dfrac{0.6}{3} = 0.2$ so take values of x every 0.2:

x	0.4	0.6	0.8	1
y	0.1787...	0.4680...	1.0601...	2.4255...

So when splitting into 3 equal strips, $\int_{0.4}^{1} \tan^2 x \, dx$
$\approx \dfrac{1}{2} \times 0.2 \times \{0.1787... + 2.4255... + 2(0.4680... + 1.0601...)\}$
$= 0.566067... = 0.566$ (3 s.f.)
[4 marks available — 1 mark for using the correct x-values, 1 mark for using the correct y-values, 1 mark for using the trapezium rule correctly, 1 mark for the correct answer]

b) $\sec^2 x = 1 + \tan^2 x$ so
$\int_{0.4}^{1} (\sec^2 x + 4) \, dx = \int_{0.4}^{1} (\tan^2 x + 5) \, dx$ *[1 mark]*
$= \int_{0.4}^{1} \tan^2 x \, dx + 5\int_{0.4}^{1} 1 \, dx = \int_{0.4}^{1} \tan^2 x \, dx + 5[x]_{0.4}^{1}$ *[1 mark]*
$\approx 0.5660... + 5(1 - 0.4) = 0.5660... + 3 = 3.57$ (3 s.f.) *[1 mark]*
[3 marks available in total — as above]

9 When x is small, $\sin x \approx x$ and $\cos x \approx 1 - \dfrac{x^2}{2}$.
$\dfrac{\text{cosec } 2x(1 - \cos 4x)}{6x} = \dfrac{1 - \cos 4x}{6x \sin 2x}$ *[1 mark]*
$\approx \dfrac{1 - \left(1 - \dfrac{(4x)^2}{2}\right)}{6x \times 2x}$ *[1 mark]* $= \dfrac{8x^2}{12x^2} = \dfrac{2}{3}$ as required *[1 mark]*
[3 marks available in total — as above]

10 a) Let $u = \ln x$ and $\dfrac{dv}{dx} = 1$. Then $\dfrac{du}{dx} = \dfrac{1}{x}$ and $v = x$ *[1 mark]*
Using integration by parts:
$\int_{1}^{4} \ln x \, dx = [x \ln x]_{1}^{4} - \int_{1}^{4} x \times \dfrac{1}{x} \, dx$ *[1 mark]*
$= [x \ln x]_{1}^{4} - \int_{1}^{4} 1 \, dx = [x \ln x - x]_{1}^{4}$ *[1 mark]*
$= (4 \ln 4 - 4) - (1 \ln 1 - 1) = 4 \ln 4 - 3$ *[1 mark]*
$= \ln 4^4 - 3 = \ln 256 - 3$ *[1 mark]*
[5 marks available in total — as above]

b) $A = \int_{1}^{4} \sqrt{x} \, dx - \int_{1}^{4} \ln x \, dx$ *[1 mark]*
$\int_{1}^{4} \sqrt{x} \, dx = \dfrac{2}{3}[x^{\frac{3}{2}}]_{1}^{4}$ *[1 mark]*
$= \dfrac{2}{3}(4^{\frac{3}{2}} - 1^{\frac{3}{2}}) = \dfrac{2}{3}(8 - 1) = \dfrac{14}{3}$ *[1 mark]*
So $A = \dfrac{14}{3} - (\ln 256 - 3) = \dfrac{23}{3} - \ln 256$ *[1 mark]*
[4 marks available in total — as above]

11 $f'\left(\dfrac{\pi}{3}\right) = \lim_{h \to 0} \dfrac{f\left(\dfrac{\pi}{3} + h\right) - f\left(\dfrac{\pi}{3}\right)}{h} = \lim_{h \to 0} \dfrac{\cos\left(\dfrac{\pi}{3} + h\right) - \cos \dfrac{\pi}{3}}{h}$ *[1 mark]*
$\cos\left(\dfrac{\pi}{3} + h\right) = \cos \dfrac{\pi}{3} \cos h - \sin \dfrac{\pi}{3} \sin h$ *[1 mark]*
$= \dfrac{1}{2} \cos h - \dfrac{\sqrt{3}}{2} \sin h$
So $f'\left(\dfrac{\pi}{3}\right) = \lim_{h \to 0} \left(\dfrac{\dfrac{1}{2}\cos h - \dfrac{\sqrt{3}}{2}\sin h - \dfrac{1}{2}}{h}\right)$ *[1 mark]*
$= \lim_{h \to 0}\left(\dfrac{1}{2}\left(\dfrac{\cos h - 1}{h}\right) - \dfrac{\sqrt{3}}{2}\left(\dfrac{\sin h}{h}\right)\right)$
$= \dfrac{1}{2}\lim_{h \to 0}\left(\dfrac{\cos h - 1}{h}\right) - \dfrac{\sqrt{3}}{2}\lim_{h \to 0}\left(\dfrac{\sin h}{h}\right)$
As $h \to 0$, $\sin h \to h$ so $\dfrac{\sin h}{h} \to \dfrac{h}{h} = 1$ *[1 mark]*
As $h \to 0$, $\cos h \to 1 - \dfrac{h^2}{2}$ so
$\dfrac{\cos h - 1}{h} \to \dfrac{\left(1 - \dfrac{h^2}{2}\right) - 1}{h} = \dfrac{-\dfrac{h^2}{2}}{h} = -\dfrac{h}{2} \to 0$ *[1 mark]*
$\Rightarrow f'\left(\dfrac{\pi}{3}\right) = \dfrac{1}{2} \times 0 - \dfrac{\sqrt{3}}{2} \times 1 = -\dfrac{\sqrt{3}}{2}$ *[1 mark]*
[6 marks available in total — as above]

12 $u = 5x^4 - 6 \Rightarrow \dfrac{du}{dx} = 20x^3$, $v = 3e^{2x} \Rightarrow \dfrac{dv}{dx} = 6e^{2x}$ *[1 mark]*
so $\dfrac{dy}{dx} = \dfrac{20x^3 \times 3e^{2x} - 6e^{2x}(5x^4 - 6)}{(3e^{2x})^2}$ *[1 mark]*
When $x = 0$, $\dfrac{dy}{dx} = \dfrac{0 - 6e^0(0 - 6)}{(3e^0)^2} = \dfrac{36}{9} = 4$ *[1 mark]*

So gradient of normal $= -\dfrac{1}{4}$ *[1 mark]*
When $x = 0$, $y = \dfrac{0 - 6}{3e^0} = -2$ *[1 mark]*
So normal is $y - (-2) = -\dfrac{1}{4}(x - 0)$
$\Rightarrow y + 2 = -\dfrac{1}{4}x \Rightarrow x + 4y + 8 = 0$ *[1 mark]*
[6 marks available in total — as above]

You could use $y = mx + c$ to get $c = -2 \Rightarrow y = -\dfrac{1}{4}x - 2$
$\Rightarrow x + 4y + 8 = 0$.

13 $C = 20 - \dfrac{9}{t+1} = 20 - 9(t+1)^{-1}$
$\Rightarrow \dfrac{dC}{dt} = 9(t+1)^{-2} = \dfrac{9}{(t+1)^2}$ *[1 mark]*
$\dfrac{dH}{dC} = \dfrac{dH}{dt} \div \dfrac{dC}{dt}$ *[1 mark]*
$= \dfrac{t^{0.2}}{10} \div \dfrac{9}{(t+1)^2} = \dfrac{t^{0.2}}{10} \times \dfrac{(t+1)^2}{9} = \dfrac{t^{0.2}(t+1)^2}{90}$
When $t = 5$, $\dfrac{dH}{dC} = \dfrac{5^{0.2}(5+1)^2}{90}$ *[1 mark]*
$= 0.5518... = 0.552$ cm °C^{-1} (3 s.f.) *[1 mark]*
[4 marks available in total — as above]

You could also have found the values of $\dfrac{dH}{dt}$ (= 0.1379...)
and $\dfrac{dC}{dt}$ (= 0.25), and divided these to find $\dfrac{dH}{dC}$.

14 For a right-angled triangle with hypotenuse c
and other sides a and b, assume that $c \geq a + b$. *[1 mark]*
Then $c^2 \geq (a + b)^2 \Rightarrow c^2 \geq a^2 + 2ab + b^2$ *[1 mark]*
Using Pythagoras' theorem, $a^2 + b^2 = c^2$ which gives
$c^2 \geq c^2 + 2ab \Rightarrow 2ab \leq 0$ *[1 mark]*. This is not possible since
a and b are lengths so must be positive and so $2ab > 0$.
Therefore the assumption that $c \geq a + b$ is incorrect,
so $c < a + b$ *[1 mark]*.
[4 marks available in total — as above]

15 a) $t = 0 \Rightarrow d = \sqrt{3} \sin 0 - \cos 0 + 5 = 0 - 1 + 5 = 4$ m *[1 mark]*

b) $R \sin\left(\dfrac{t}{2} - \alpha\right) = R \sin \dfrac{t}{2} \cos \alpha - R \cos \dfrac{t}{2} \sin \alpha$
$= \sqrt{3} \sin \dfrac{t}{2} - \cos \dfrac{t}{2} \Rightarrow R \sin \alpha = 1$ and $R \cos \alpha = \sqrt{3}$
$(R \sin \alpha)^2 + (R \cos \alpha)^2 = 1^2 + (\sqrt{3})^2$
$\Rightarrow R^2(\sin^2 \alpha + \cos^2 \alpha) = 1 + 3 \Rightarrow R^2 = 4 \Rightarrow R = 2$ *[1 mark]*
$\dfrac{R \sin \alpha}{R \cos \alpha} = \tan \alpha = \dfrac{1}{\sqrt{3}}$ *[1 mark]* $\Rightarrow \alpha = \dfrac{\pi}{6}$ *[1 mark]*
So $\sqrt{3} \sin \dfrac{t}{2} - \cos \dfrac{t}{2} = 2 \sin\left(\dfrac{t}{2} - \dfrac{\pi}{6}\right)$
[3 marks available in total — as above]

c) $d = 2 \sin\left(\dfrac{t}{2} - \dfrac{\pi}{6}\right) + 5$ so $d_{\min} = 2(-1) + 5 = 3$ m *[1 mark]*
This occurs when $\sin\left(\dfrac{t}{2} - \dfrac{\pi}{6}\right) = -1$
$\Rightarrow \dfrac{t}{2} - \dfrac{\pi}{6} = \dfrac{3\pi}{2}$ *[1 mark]*
$\Rightarrow \dfrac{t}{2} = \dfrac{5\pi}{3} \Rightarrow t = \dfrac{10\pi}{3} = 10.4719...$ hours
≈ 10 hours 28 mins (or 628 minutes) *[1 mark]*
[3 marks available in total — as above]

d) Given rate is per day so k is $\dfrac{1.2 \times 10^{-5}}{24} = 5 \times 10^{-7}$
or $\dfrac{1}{2\,000\,000}$ metres per hour. *[1 mark]*

16 a) To avoid division by zero $3x - 1 \neq 0 \Rightarrow x \neq \dfrac{1}{3}$ *[1 mark]*

b) $-1 \leq \cos x \leq 1 \Rightarrow -2 \leq 2\cos x \leq 2$
$\Rightarrow -2 + \dfrac{1}{3} \leq 2\cos x + \dfrac{1}{3} \leq 2 + \dfrac{1}{3} \Rightarrow -\dfrac{5}{3} \leq g(x) \leq \dfrac{7}{3}$
[2 marks available — 1 mark for each correct bound]

c) $fg(x) = f(g(x)) = \dfrac{1}{\left(3\left(2\cos x + \dfrac{1}{3}\right) - 1\right)^2}$ *[1 mark]*
$= \dfrac{1}{(6\cos x + 1 - 1)^2} = \dfrac{1}{36 \cos^2 x} = \dfrac{1}{36} \sec^2 x$ *[1 mark]*
$\int (fg(x) + 6f(x)) \, dx = \dfrac{1}{36} \int \sec^2 x \, dx + 6 \int \dfrac{1}{(3x-1)^2} \, dx$
$= \dfrac{1}{36} \tan x$ *[1 mark]* $+ 6(-1)\left(\dfrac{1}{3}\right)\dfrac{1}{(3x-1)} + C$ *[1 mark]*
$\Rightarrow \dfrac{1}{36} \tan x - \dfrac{2}{(3x-1)} + C$ *[1 mark]*
[5 marks available in total – as above]

Set 1 Paper 3 — Statistics and Mechanics

1. a) The probability of any exact value is zero for a continuous distribution, so $a = 0$. *[1 mark]*

   b) The normal distribution is symmetrical about the mean, so $b = 0.5$ or $\frac{1}{2}$. *[1 mark]*

   c) $P(X < \mu + c\sigma) = 0.8 \Rightarrow P\left(Z < \frac{(\mu + c\sigma) - \mu}{\sigma}\right) = 0.8$ *[1 mark]*
   $\Rightarrow P(Z < c) = 0.8 \Rightarrow c = \Phi^{-1}(0.8) = 0.842$ (3 s.f.) *[1 mark]*
   [2 marks available in total — as above]

2. a) E.g. He could pick a random starting point (1st, 2nd or 3rd) then take every third reading until he has 10. *[1 mark]*

   b) Mean $= \frac{\sum x}{n} = \frac{207}{10} = 20.7$ *[1 mark]*
   S.d. $= \sqrt{\frac{\sum x^2}{n} - \bar{x}^2} = \sqrt{\frac{4415}{10} - 20.7^2} = 3.606...$ *[1 mark]*
   So any outliers are outside
   $(20.7 - 2.5 \times 3.606...) \le x \le (20.7 + 2.5 \times 3.606...)$
   $\Rightarrow 11.682... \le x \le 29.717...$ *[1 mark]*
   None of the data is outside this interval and therefore there are no outliers. *[1 mark]*
   [4 marks available in total — as above]

   c) In order, Rajad's data is: 14 18 18 19 20 20 23 24 24 27
   E.g. The median of the 2015 data is 20 kn, and the median of the 1987 data is 18.5 kn. This suggests that the average daily maximum wind gust at Heathrow in 2015 is slightly higher than in 1987.
   The interquartile range of the 2015 data is $24 - 18 = 6$. This is slightly larger than the 1987 IQR of $21 - 16 = 5$, which suggests that the daily maximum wind gust at Heathrow in 2015 was less consistent than in 1987.
   [4 marks available – 1 mark for calculation of the median of Rajad's data, 1 mark for suitable comparison based on medians, 1 mark for calculation of a suitable measure of variation of Rajad's data, 1 mark for suitable comparison based on this measure of variation]
   Alternatively, you could compare the range of the two data sets. You could also argue that the central tendency and variation for both distributions are actually quite similar which is evidence that there are no significant differences.

   d) E.g. You would expect the data to be higher, since Camborne is on the coast so it's likely to be windier.
   [1 mark for a sensible conclusion based on locations of Heathrow and Camborne]

3. a) E.g. He has assumed: that he always plays all 18 holes / that the probability of scoring above par does not vary from hole to hole / that scoring above par on any hole is independent from his score on other holes / that he doesn't get any better or worse over time. *[1 mark for a valid assumption]*

   b) $H_0: p = 0.6$, $H_1: p \ne 0.6$ *[1 mark]*
   This is a two-tailed test, so the critical region is $H \le a$ and $H \ge b$, where $P(H \le a) \le 0.05$ and $P(H \ge b) \le 0.05$ *[1 mark]*
   From calculator: $P(H \le 6) = 0.0202...$
   $P(H \le 7) = 0.0576...$ so $a = 6$
   $P(H \ge 14) = 1 - P(H \le 13) = 1 - 0.9058... = 0.0941...$
   $P(H \ge 15) = 1 - P(H \le 14) = 1 - 0.9672... = 0.0327...$ so $b = 15$
   [1 mark for a correct method, 1 mark for the correct value of either a or b]
   So the critical region is $H \le 6$ and $H \ge 15$. *[1 mark]*
   [5 marks available in total — as above]

4. a) Substituting $c = P - 1000$ into $t = 35 - 1.25c$
   $\Rightarrow t = 35 - 1.25(P - 1000)$
   $\Rightarrow t = 35 - 1.25P + 1250 \Rightarrow t = 1285 - 1.25P$ *[1 mark]*

   b) $r = -0.786$ indicates a (strong) negative correlation, which suggests that as pressure increases, rainfall totals decrease. *[1 mark]*
   You could say that as pressure decreases, rainfall totals increase.

   c) The line of best fit has equation $y = 5.9 - 0.2x$
   Since ($\ln t$) is on the vertical axis and $(P - 1000)$ is on the horizontal axis, this gives: $\ln t = 5.9 - 0.2(P - 1000)$ *[1 mark]*
   Rearrange into the form $t = kb^{(P-1000)}$: $e^{\ln t} = e^{5.9 - 0.2(P-1000)}$
   $\Rightarrow t = e^{5.9}e^{-0.2(P-1000)} \Rightarrow t = e^{5.9}(e^{-0.2})^{(P-1000)}$ *[1 mark]*
   So $k = e^{5.9} = 365.03... = 365$ (3 s.f.)
   and $b = e^{-0.2} = 0.81873... = 0.819$ (3 s.f.) *[1 mark for both]*
   [3 marks available in total — as above]
   You could also rearrange $t = kb^{(P-1000)}$ into the form $\ln t = \ln k + (\ln b)(P - 1000)$ to get $\ln k = 5.9$ and $\ln b = -0.2$

5. a) $X \sim B(20, 0.05)$, so using the binomial tables:
   $P(X > 3) = 1 - P(X \le 3) = 1 - 0.9841 = 0.0159$
   [2 marks available – 1 mark for a correct method, 1 mark for the correct final probability]

   b) If 16 or more patients do not have the condition, then 4 or fewer patients have the condition, so the required probability is $P(X \le 4)$. *[1 mark]*
   From the tables: $P(X \le 4) = 0.9974$ *[1 mark]*
   [2 marks available in total — as above]
   You could also find $P(Y \ge 16) = 1 - P(Y \le 15)$, where Y is the number of patients who do not have the condition, and $Y \sim B(20, 0.95)$

   c) $n = 5 \times 20 = 100$, so: $C \sim B(100, 0.05)$
   $\Rightarrow \mu = 100 \times 0.05 = 5$
   and $\sigma^2 = 100 \times 0.05 \times (1 - 0.05) = 4.75$
   $\Rightarrow Y \sim N(5, 4.75)$ *[1 mark]*
   $P(C \le 10) \approx P(Y \le 10.5)$ *[1 mark]*
   $= 0.99419... = 0.994$ (3 s.f.) *[1 mark]*
   [3 marks available in total — as above]

   d) E.g. p is very far from 0.5, so the approximation might not be very good / he is using a continuous distribution to model a discrete variable. *[1 mark]*

6. a) If A and B are mutually exclusive, $P(A \cap B) = 0$ *[1 mark]*

   b) If A and B are independent, $P(A \cap B) = P(A) \times P(B)$
   $= 0.7 \times 0.15 = 0.105$ or $\frac{21}{200}$ *[1 mark]*
   $P(A \cup B) = P(A) + P(B) - P(A \cap B)$
   $= 0.7 + 0.15 - 0.105 = 0.745$ or $\frac{149}{200}$ *[1 mark]*
   [2 marks available in total — as above]

   c) i) $P(A' \cap B) = P(B) - P(A \cap B)$
   $= 0.15 - 0.037 = 0.113$ or $\frac{113}{1000}$ *[1 mark]*

   ii) $P(A' | B') = \frac{P(A' \cap B')}{P(B')}$
   $P(A' \cap B') = 1 - P(A \cup B) = 1 - (P(A) + P(B) - P(A \cap B))$
   $= 1 - (0.7 + 0.15 - 0.037) = 0.187$ *[1 mark]*
   So $P(A' | B') = \frac{0.187}{1 - 0.15} = \frac{0.187}{0.85} = 0.22$ or $\frac{11}{50}$ *[1 mark]*
   [2 marks available in total — as above]
   You might find it useful to draw a Venn diagram here.

7. a) P(first is in $130 \le t < 135$ class, second is not)
   $= \frac{20}{120} \times \frac{100}{119} = \frac{2000}{14\,280} = \frac{50}{357}$ *[1 mark]*
   P(first is not in $130 \le t < 135$ class, second is)
   $= \frac{100}{120} \times \frac{20}{119} = \frac{2000}{14\,280} = \frac{50}{357}$
   So P(exactly one is in $130 \le t < 135$ class)
   $= \frac{50}{357} + \frac{50}{357} = \frac{100}{357}$ or 0.280 (3 s.f.) *[1 mark]*
   [2 marks available in total — as above]
   Drawing a tree diagram might help you to find these probabilities.

   b) The $140 \le t < 145$ class will have height x cm,
   area $= \frac{A}{120} \times 37 = \frac{37A}{120}$ cm² and width $= \frac{37A}{120} \div x = \frac{37A}{120x}$ cm.
   The $150 \le t < 170$ class is 4 times as wide so will have
   width $= \frac{37A}{120x} \times 4 = \frac{37A}{30x}$ cm, area $= \frac{A}{120} \times 10 = \frac{A}{12}$ cm²
   and height $= \frac{A}{12} \div \frac{37A}{30x} = \frac{A}{12} \times \frac{30x}{37A} = \frac{5x}{74}$ cm.
   [3 marks available — 1 mark for a suitable method, 1 mark for the correct width, 1 mark for the correct height]

c) Using your calculator, mean ≈ 141.833... *[1 mark]*
$T \sim N(144, 7^2)$ and $H_0: \mu = 144$, $H_1: \mu < 144$ *[1 mark]*
Test statistic = $\dfrac{141.833... - 144}{7/\sqrt{120}} = -3.390...$ *[1 mark]*
$a = 2.5\% = 0.025$, so from the percentage points tables:
Critical value = 1.9600 *[1 mark]*
3.390... > 1.9600 so the result is significant. *[1 mark]*
There is enough evidence at the 2.5% significance level to reject H_0 and support the claim that the mean of T has decreased. *[1 mark]*
[6 marks available in total — as above]

8 a) The velocity is equal to the area under the graph.
Using the formula for the area of a trapezium:
$57.8 = \dfrac{1}{2}(4.2 + 9.4)b$ *[1 mark]*
$\Rightarrow 57.8 = 6.8b \Rightarrow b = 8.5$ ms^{-2} as required. *[1 mark]*
[2 marks available in total — as above]

b) $s = s$, $u = 57.8$ ms^{-1}, $v = 10$ ms^{-1}, $a = -12.5$ ms^{-2}, $t = t$
Using $v = u + at$: $10 = 57.8 - 12.5t$ *[1 mark]*
$\Rightarrow 12.5t = 47.8 \Rightarrow t = 3.824 = 3.8$ seconds (2 s.f.) *[1 mark]*
Using $v^2 = u^2 + 2as$: $10^2 = 57.8^2 + 2(-12.5)s$ *[1 mark]*
$\Rightarrow 25s = 57.8^2 - 10^2 = 3240.84$
$\Rightarrow s = 129.6336 = 130$ m (2 s.f.) *[1 mark]*
[4 marks available in total — as above]
You could have used different suvat equations here or found the values by using an acceleration-time graph.

c) $\mathbf{v} = \dfrac{d\mathbf{s}}{dt} = \left(\dfrac{3t^2}{15}\right)\mathbf{i} + (1)\mathbf{j} = \left(\left(\dfrac{t^2}{5}\right)\mathbf{i} + \mathbf{j}\right)$ ms^{-1} *[1 mark]*
At $t = 9.4$, $\mathbf{v} = \dfrac{9.4^2}{5}\mathbf{i} + \mathbf{j} = (17.672\mathbf{i} + \mathbf{j})$ ms^{-1} *[1 mark]*
So speed = $\sqrt{17.672^2 + 1^2}$ *[1 mark]*
= $\sqrt{313.299...}$
= 17.7002... = 17.7 ms^{-1} (3 s.f.) *[1 mark]*
[4 marks available in total — as above]

d) Acceleration is not constant ($\mathbf{a} = \dfrac{d\mathbf{v}}{dt} = \dfrac{6t}{15}\mathbf{i} + 0 = \dfrac{2}{5}t\mathbf{i}$), so the kinematic equations cannot be used to model the motion of the drone. *[1 mark]*

9 a) Using $F_{net} = ma$ for the whole system: *[1 mark]*
$4900 - B - 4B - T + T = (750 + 1250) \times 1.5$ *[1 mark]*
$\Rightarrow 4900 - 5B = 3000 \Rightarrow 5B = 1900 \Rightarrow B = 380$ N *[1 mark]*
[3 marks available in total — as above]

b) Using $F_{net} = ma$ for the trailer:
$T - 4B = 750 \times 1.5$ *[1 mark]* $\Rightarrow T - (4 \times 380) = 1125$
$\Rightarrow T = 1125 + 1520 = 2645$ N *[1 mark]*
[2 marks available in total — as above]
You could also resolve forces on the car — you'd get the same answer.

c) The chain is light (has zero/negligible mass/weight, etc.) *[1 mark]*

10 Since $\tan \alpha = \dfrac{5}{12}$, hypotenuse = $\sqrt{5^2 + 12^2} = 13$ so:
$\sin \alpha = \dfrac{5}{13}$ and $\cos \alpha = \dfrac{12}{13}$ *[1 mark]*
Resolving forces perpendicular to the plane:
$R = 325 \cos \alpha = 325 \times \dfrac{12}{13} = 300$ N *[1 mark]*
The object is on the point of sliding, so $F = \mu R = 300\mu$ *[1 mark]*
Resolving forces parallel to the plane:
$191 = F + 325 \sin \alpha$ *[1 mark]*
$= 300\mu + \left(325 \times \dfrac{5}{13}\right) = 300\mu + 125$ *[1 mark]*
$\Rightarrow 300\mu = 66 \Rightarrow \mu = \dfrac{66}{300} = 0.22$ or $\dfrac{11}{50}$ *[1 mark]*
[6 marks available in total — as above]

11 a) Let R be the reaction force at A.
Resolving forces vertically: $R + \dfrac{2}{3}R = 52g + 86g$ *[1 mark]*
$\Rightarrow \dfrac{5}{3}R = 138g \Rightarrow R = 82.8g$ N as required. *[1 mark]*
[2 marks available in total — as above]

b) E.g. Equating moments about A:
$(52g \times AX) + (86g \times 3) = \left(\dfrac{2}{3}(82.8g) \times 6\right)$
$52gAX + 258g = 331.2g$
$52AX = 73.2$
$AX = 1.4076...$ m = 141 cm (to the nearest cm)
[3 marks available — 1 mark for equating moments, 1 mark for a correct equation, 1 mark for the correct answer]
You could have taken moments about any point, not just A.

c) i) The child's weight force acts at the point X. *[1 mark]*
ii) E.g. The plank remains straight/the same shape so all forces are perpendicular/all distances are parallel to the line AB. *[1 mark]*

12 a) Horizontally, $u_H = v_H = (225 \cos 3°)$ ms^{-1} and $s = 1900$ m
Using time = displacement ÷ speed
$t = \dfrac{1900}{225 \cos 3°} = 8.456...$ seconds *[1 mark]*
Vertically, taking upwards as positive:
$s = s$, $u_V = (225 \sin 3°)$ ms^{-1}, $a = -9.81$ ms^{-2}, $t = 8.456...$ s
Using $s = ut + \dfrac{1}{2}at^2$:
$s = (225 \sin 3°) \times 8.456... + \dfrac{1}{2}(-9.81)(8.456...)^2$ *[1 mark]*
= $-251.15...$ so the height is 251 m (3 s.f.) *[1 mark]*
[3 marks available in total — as above]

b) Just before it hits the water, $t = 8.456...$ from part a).
Horizontally: $v_H = 225 \cos 3° = 224.69...$ ms^{-1}
Vertically: $v = u + at \Rightarrow v_V = 225 \sin 3° - 9.81(8.456...)$
= $-71.178...$ ms^{-1} *[1 mark]*
So speed = $\sqrt{(71.178...)^2 + (224.69...)^2}$ *[1 mark]*
= 235.69... = 236 ms^{-1} (3 s.f.) *[1 mark]*
[3 marks available in total — as above]

c) Vertically, $s = 0$, so using $s = ut + \dfrac{1}{2}at^2$:
$0 = (225 \sin 3°)t + \dfrac{1}{2}(-9.81)t^2$ *[1 mark]*
$\Rightarrow t(225 \sin 3° - 4.905t) = 0$
$\Rightarrow t = 0$ (ignore) or $4.905t = 225 \sin 3°$
$\Rightarrow t = 2.400... = 2.40$ seconds (3 s.f.) *[1 mark]*
[2 marks available in total — as above]

13 a) The playing surface must be smooth (or have no friction/resistance). *[1 mark]*

b) E.g. $\tan^{-1}\left(\dfrac{6}{4}\right) = 56.309...°$ *[1 mark]*
So bearing = $270° + 56.309...° \approx 326°$ *[1 mark]*
[2 marks available in total — as above]

c) The position vector of the puck at time t is:
$(30\mathbf{i} + 13\mathbf{j}) + (-4\mathbf{i} + 6\mathbf{j})t = (30 - 4t)\mathbf{i} + (13 + 6t)\mathbf{j}$ *[1 mark]*
Cindy's displacement at time t can be found using
$\mathbf{s} = \mathbf{u}t + \dfrac{1}{2}\mathbf{a}t^2$ where $\mathbf{u} = 0$ ms^{-1} and $\mathbf{a} = -a\mathbf{i}$ ms^{-1}
$\Rightarrow \mathbf{s} = \left(-\dfrac{1}{2}at^2\right)\mathbf{i}$ m *[1 mark]*
So her position at time t is:
$(34\mathbf{i} + 22\mathbf{j}) - \left(\dfrac{1}{2}at^2\right)\mathbf{i} = \left(34 - \dfrac{1}{2}at^2\right)\mathbf{i} + 22\mathbf{j}$ *[1 mark]*
Equating $\mathbf{j}$ components:
$13 + 6t = 22 \Rightarrow 6t = 9 \Rightarrow t = 1.5$ seconds *[1 mark]*
Equating $\mathbf{i}$ components and substituting in $t = 1.5$:
$30 - 4(1.5) = 34 - \dfrac{1}{2}a(1.5^2) \Rightarrow 24 = 34 - 1.125a$
$\Rightarrow a = 8.8888... = 8.89$ ms^{-2} (3 s.f.) *[1 mark]*
[5 marks available in total — as above]

d) The position vector of the puck at time t is:
$(15\mathbf{i} + 15\mathbf{j}) + (-44\mathbf{i} - A\mathbf{j})t = (15 - 44t)\mathbf{i} + (15 - At)\mathbf{j}$ *[1 mark]*
The puck reaches the goal when the $\mathbf{i}$-component is 4:
$15 - 44t = 4 \Rightarrow 44t = 11 \Rightarrow t = 0.25$ seconds *[1 mark]*
At this point, the $\mathbf{j}$-component is: $15 - 0.25A$
The goalkeeper blocks the goal between $(4\mathbf{i} + 12\mathbf{j})$ and $(4\mathbf{i} + 13.3\mathbf{j})$, so in order to score:
$13.3 < 15 - 0.25A < 14$ *[1 mark]*
$\Rightarrow 1 < 0.25A < 1.7 \Rightarrow 4 < A < 6.8$ *[1 mark]*
[4 marks available in total — as above]

Set 2 Paper 1 — Pure Mathematics 1

1. a) $b^2 - 4ac = 12^2 - 4(3)(19) = 144 - 228 = -84$ *[1 mark]*
   The discriminant is negative, so the graph does not intersect the x-axis. *[1 mark]*
   [2 marks available in total — as above]

   b) $3x^2 + 12x + 19 = 3(x^2 + 4x) + 19$ *[1 mark]*
   $= 3((x + 2)^2 - 4) + 19$ *[1 mark]* $= 3(x + 2)^2 + 7$ *[1 mark]*
   [3 marks available in total — as above]

   c)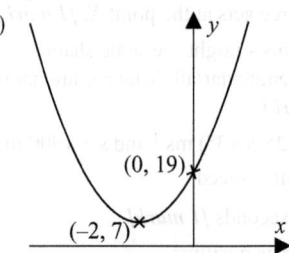

   [3 marks available — 1 mark for correct shape with minimum in the correct quadrant, 1 mark for labelling the stationary point (–2, 7), 1 mark for labelling the y-intercept (0, 19)]

   d) A translation 2 units right and 7 units down or by vector $\begin{pmatrix} 2 \\ -7 \end{pmatrix}$ *[1 mark]*, then a vertical stretch by scale factor $\frac{1}{3}$ *[1 mark]*.
   [2 marks available in total — as above]
   Alternatively, a vertical stretch by scale factor $\frac{1}{3}$ followed by a translation 2 units right and $\frac{7}{3}$ units down.

2. If w is the width then the length is $2w$.
   The perimeter can be up to 40 m,
   so $w + 2w + w + 2w \leq 40 \Rightarrow 6w \leq 40 \Rightarrow w \leq \frac{20}{3}$ *[1 mark]*.
   The area needs to be at least 60 m^2,
   so $w \times 2w \geq 60 \Rightarrow 2w^2 \geq 60 \Rightarrow w^2 \geq 30$
   $\Rightarrow w \leq -\sqrt{30}$ or $w \geq \sqrt{30}$ *[1 mark]*
   Ignoring negative values (as the width can't be negative)
   gives $\sqrt{30} \leq w \leq \frac{20}{3}$. So the difference between the maximum and minimum width is:
   $\frac{20}{3} - \sqrt{30}$ *[1 mark]* $= \frac{20 - 3\sqrt{30}}{3} = \frac{20 - \sqrt{270}}{3}$ m *[1 mark]*
   [4 marks available in total — as above]

3. a) $\frac{dy}{dx} = 2e^{2x}$ *[1 mark]*
   a is directly proportional to $b \Rightarrow a = kb$ for some constant k.
   Here, $y = \frac{1}{2}\frac{dy}{dx}$, so y is directly proportional to $\frac{dy}{dx}$. *[1 mark]*
   [2 marks available in total — as above]

   b) At $(2, e^4)$, $\frac{dy}{dx} = 2e^4$. Using $y - y_1 = m(x - x_1)$ gives
   $y - e^4 = 2e^4(x - 2)$ *[1 mark]* $\Rightarrow y = 2e^4x - 3e^4$
   The line crosses the x-axis at $y = 0$, so $0 = 2e^4x - 3e^4$
   $\Rightarrow x = \frac{3}{2}$ so the coordinates are $(\frac{3}{2}, 0)$ *[1 mark]*
   [2 marks available in total — as above]

4. a) Freya is incorrect *[1 mark]*. Although the value of $\sin x$ does repeat every 2π, the value of x^2 does not, so she needs to show it for all values of x. *[1 mark]*
   [2 marks available in total — as above]

   b) For all values of x, $3 \sin x \geq -3$ so $3 \sin x + 3 \geq 0$
   When $x \neq 0$, $x^2 > 0$ for all values of x.
   So $3 \sin x + x^2 + 3 > 0$ when $x \neq 0$
   When $x = 0$, $3 \sin 0 + 0^2 + 3 = 3$ so
   $3 \sin x + x^2 + 3 > 0$ for all values of x.
   [3 marks available — 1 mark for stating that $3 \sin x \geq -3$, 1 mark for arguing that x^2 is positive for non-zero values, 1 mark for a fully correct proof]

5. $f(-2) = 0 \Rightarrow (-2)^3 - b(-2)^2 + 2(-2) + 40 = 0$ *[1 mark]*
   $\Rightarrow -8 - 4b - 4 + 40 = 0 \Rightarrow 4b = 28 \Rightarrow b = 7$ *[1 mark]*
   It cuts the x-axis at -2 so $(x + 2)$ is a factor *[1 mark]*

Dividing the cubic expression by $(x + 2)$ gives:

$$\begin{array}{r} x^2 - 9x + 20 \\ x + 2 \overline{\smash{)}x^3 - 7x^2 + 2x + 40} \\ \underline{x^3 + 2x^2} \\ -9x^2 + 2x \\ \underline{-9x^2 - 18x} \\ 20x + 40 \\ \underline{20x + 40} \\ 0 \end{array}$$ *[1 mark for a correct method]*

$f(x) = (x + 2)(x^2 - 9x + 20)$ *[1 mark]* $= (x + 2)(x - 4)(x - 5)$ *[1 mark]*
[6 marks available in total — as above]
You could have used an alternative method to take a factor of (x + 2) out of the expression.

6. a) The formula for a geometric sequence is $u_n = ar^{n-1}$
   The ball is dropped from 6 m so $a = 6$.
   After the first bounce, $u_2 = 5.52$ m so:
   $u_2 = a \times r^1 \Rightarrow 5.52 = 6 \times r \Rightarrow r = 0.92 \Rightarrow u_n = 6 \times 0.92^{n-1}$
   When the maximum height for a bounce is less than 1 m:
   $u_n < 1 \Rightarrow 6 \times 0.92^{n-1} < 1 \Rightarrow 0.92^{n-1} < \frac{1}{6}$
   $\Rightarrow \ln 0.92^{n-1} < \ln \frac{1}{6} \Rightarrow (n - 1)\ln 0.92 < \ln \frac{1}{6}$
   $\Rightarrow n - 1 > \frac{\ln \frac{1}{6}}{\ln 0.92} \Rightarrow n - 1 > 21.4... \Rightarrow n > 22.4...$
   So a maximum height of less than 1 metre is first achieved when $n = 23$, which is the 23rd maximum height, so after the 22nd bounce.
   [5 marks available — 1 mark for finding the correct value of r, 1 mark for setting up the formula for a geometric sequence, 1 mark for using logs to solve the inequality, 1 mark for solving the inequality correctly, 1 mark for interpreting the answer in the context of the question]

   b) $S_\infty = \frac{a}{1 - r} = \frac{6}{1 - 0.92} = \frac{6}{0.08} = 75$ m *[1 mark]*
   After each bounce the ball goes up and down, so you need to double S_∞, but it only travels the initial height (6 m) once, so:
   Total distance travelled = $(2 \times 75) - 6 = 144$ m *[1 mark]*
   [2 marks available in total — as above]

7.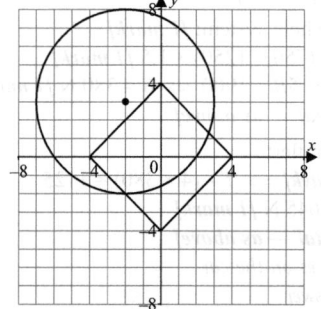

   From the diagram you can see that the circle intersects the square twice, and that one of the points of intersection is $(-2, -2)$ *[1 mark]*.
   To find the other point of intersection, in the top-right quadrant, solve the equations simultaneously.
   The circle has equation $(x + 2)^2 + (y - 3)^2 = 5^2$ *[1 mark]*
   The equation of the line which makes up the square where $x > 0$ and $y > 0$ has gradient -1 and y-intercept 4,
   so the equation is $y = -x + 4$ *[1 mark]*
   Substitute $y = -x + 4$ into $(x + 2)^2 + (y - 3)^2 = 25$ to give
   $(x + 2)^2 + ((-x + 4) - 3)^2 = 25$ *[1 mark]*
   $\Rightarrow (x + 2)^2 + (-x + 1)^2 = 25 \Rightarrow (x^2 + 4x + 4) + (x^2 - 2x + 1) = 25$
   $\Rightarrow 2x^2 + 2x - 20 = 0 \Rightarrow x^2 + x - 10 = 0$ *[1 mark]*
   $\Rightarrow x = \frac{-1 \pm \sqrt{1^2 + 40}}{2} \Rightarrow x = \frac{-1 \pm \sqrt{41}}{2}$
   Ignore the negative root since $x > 0$.
   $\Rightarrow x = \frac{-1 + \sqrt{41}}{2} = 2.701...$ *[1 mark]*
   At $x = 2.701...$, $y = -x + 4 = -2.701... + 4 = 1.298...$
   So the coordinates of the other point of intersection are $(2.70, 1.30)$ (2 d.p.) *[1 mark]*
   [7 marks available in total — as above]

8 a) $f^{-1}(x) = \frac{2x-5}{x} \Rightarrow x = \frac{2y-5}{y} \Rightarrow yx = 2y - 5$
$\Rightarrow yx - 2y = -5 \Rightarrow y(x-2) = -5 \Rightarrow y = \frac{-5}{x-2}$
$\Rightarrow f(x) = \frac{-5}{x-2}$ or $\frac{5}{2-x}$
So $fg(x) = \frac{-5}{\sqrt{2x-k}-2}$ or $\frac{5}{2-\sqrt{2x-k}}$
$fg(x)$ is undefined when $2x - k < 0 \Rightarrow x < \frac{k}{2}$
or $\sqrt{2x-k} - 2 = 0 \Rightarrow 2x - k = 4 \Rightarrow x = 2 + \frac{k}{2}$
So $fg(x)$ has domain $x \geq \frac{k}{2}$ and $x \neq 2 + \frac{k}{2}$
[3 marks available — 1 mark for finding f(x), 1 mark for finding fg(x), 1 mark for identifying a suitable domain]

b) $g(10) = \sqrt{20-k}$ and $gg(10) = \sqrt{2\sqrt{20-k} - k} = 2$ *[1 mark]*
$\Rightarrow 2\sqrt{20-k} - k = 4 \Rightarrow \sqrt{20-k} = \frac{4+k}{2}$
$\Rightarrow 20 - k = \left(\frac{4+k}{2}\right)^2 \Rightarrow 20 - k = \frac{16 + 8k + k^2}{4}$
$\Rightarrow 80 - 4k = 16 + 8k + k^2$ *[1 mark]*
$\Rightarrow k^2 + 12k - 64 = 0 \Rightarrow (k+16)(k-4) = 0$
$\Rightarrow k = 4$ *[1 mark]* (ignore $k = -16$ as k is positive)
[3 marks available in total — as above]

9 a) $6y\frac{dy}{dx} - 4\frac{dy}{dx} = -6x^2$
$\Rightarrow \frac{dy}{dx}(6y - 4) = -6x^2 \Rightarrow \frac{dy}{dx} = \frac{-3x^2}{3y-2}$ or $\frac{3x^2}{2-3y}$
[3 marks available — 1 mark for differentiating $3y^2$ correctly, 1 mark for differentiating the other terms correctly, 1 mark for the correct answer]

b) At stationary points, $\frac{dy}{dx} = 0 \Rightarrow \frac{-3x^2}{3y-2} = 0 \Rightarrow x = 0$ *[1 mark]*
At $x = 0$, $3y^2 - 4y = 4 \Rightarrow 3y^2 - 4y - 4 = 0$ *[1 mark]*
$\Rightarrow (3y+2)(y-2) = 0 \Rightarrow y = -\frac{2}{3}$ or $y = 2$
So the distance between the stationary points is:
$2 - \left(-\frac{2}{3}\right) = 2\frac{2}{3}$ or $\frac{8}{3}$ *[1 mark]*
[3 marks available in total — as above]

10 $x = \frac{2}{3}\sin\theta \Rightarrow \frac{dx}{d\theta} = \frac{2}{3}\cos\theta$ *[1 mark]* $\Rightarrow dx = \frac{2}{3}\cos\theta\, d\theta$
So $\int \frac{1}{\sqrt{4-9x^2}}\,dx = \int \frac{1}{\sqrt{4 - 9 \times \frac{4}{9}\sin^2\theta}} \cdot \frac{2}{3}\cos\theta\, d\theta$ *[1 mark]*
$= \frac{2}{3}\int \frac{\cos\theta}{\sqrt{4-4\sin^2\theta}}\,d\theta = \frac{2}{3}\int \frac{\cos\theta}{\sqrt{4}\sqrt{1-\sin^2\theta}}\,d\theta = \frac{2}{6}\int \frac{\cos\theta}{\sqrt{\cos^2\theta}}\,d\theta$
$= \frac{1}{3}\int \frac{\cos\theta}{\cos\theta}\,d\theta = \frac{1}{3}\int 1\,d\theta$ *[1 mark]* $= \frac{\theta}{3} + C$ *[1 mark]*
Rearranging $x = \frac{2}{3}\sin\theta$ gives $\theta = \arcsin\frac{3}{2}x$,
so $\int \frac{1}{\sqrt{4-9x^2}}\,dx = \frac{1}{3}\arcsin\frac{3}{2}x + C$ *[1 mark]*
[5 marks available in total — as above]

11 a) $R\sin(\theta + \alpha) = 5\sin\theta + 7\cos\theta$
$\Rightarrow R\sin\theta\cos\alpha + R\cos\theta\sin\alpha = 5\sin\theta + 7\cos\theta$
$\Rightarrow R\cos\alpha = 5$ and $R\sin\alpha = 7$
$\frac{R\sin\alpha}{R\cos\alpha} = \frac{7}{5} \Rightarrow \tan\alpha = 1.4$ *[1 mark]*
$\Rightarrow \alpha = 54.46...° = 54°$ (nearest whole degree) *[1 mark]*
$(R\sin\alpha)^2 + (R\cos\alpha)^2 = 5^2 + 7^2 \Rightarrow R^2(\sin^2\alpha + \cos^2\alpha) = 25 + 49$
$\Rightarrow R^2 = 74 \Rightarrow R = \sqrt{74}$ *[1 mark]*
[3 marks available in total — as above]

b)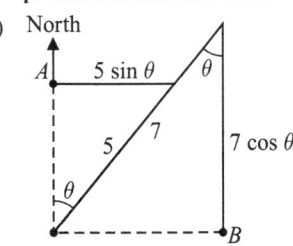
Boat A sails a distance of $5 + 5\sin\theta$
Boat B sails a distance of $7 + 7\cos\theta$ *[1 mark for both]*

$12 + 5\sin\theta + 7\cos\theta = 18 \Rightarrow 5\sin\theta + 7\cos\theta = 6$
$\sqrt{74}\sin(\theta + 54.46...°) = 6$ *[1 mark]*
$\Rightarrow \sin(\theta + 54.46...°) = \frac{6}{\sqrt{74}}$
$\Rightarrow \theta + 54.46...° = 44.22...°, 135.77...°$
$\Rightarrow \theta = -10.24...°, 81.31...°$ *[1 mark]*
You know that $0° < \theta < 90° \Rightarrow \theta = 81.31...°$, so the boats sail on a bearing of $081°$ (nearest whole degree) *[1 mark]*.
[4 marks available in total — as above]

12 a) After t years, where A is the original sum and P is the value of the investment at that time: $P = A\left(\frac{100+r}{100}\right)^t$
When the investment has doubled, $P = 2A$ and $t = T$; substituting in gives: $2A = A\left(\frac{100+r}{100}\right)^T$
$\Rightarrow 2 = \left(\frac{100+r}{100}\right)^T \Rightarrow \log 2 = \log\left(\frac{100+r}{100}\right)^T$
$\Rightarrow \log 2 = T\log\left(\frac{100+r}{100}\right)$
$\Rightarrow T = \frac{\log 2}{\log\left(\frac{100+r}{100}\right)} = \log_{\left(\frac{100+r}{100}\right)} 2$ or $\log_{(1+0.01r)} 2$
[3 marks available — 1 mark for a correct equation relating T and r, 1 mark for rearranging and taking logs, 1 mark for the correct answer]

b) $15 = \log_{\left(\frac{100+r}{100}\right)} 2 \Rightarrow \left(\frac{100+r}{100}\right)^{15} = 2$ *[1 mark]*
$\sqrt[15]{2} = \frac{100+r}{100} \Rightarrow 1.0472... = \frac{100+r}{100}$
$\Rightarrow 100 + r = 104.729...$
$\Rightarrow r = 4.729...\% = 4.73\%$ (2 d.p.) *[1 mark]*
[2 marks available in total — as above]

c) From a), compound interest will double when $t = \log_{\left(\frac{100+q}{100}\right)} 2$.
The simple interest account doubles after $2t$ years:
If the initial investment was A, then
$2A = A \times \left(\frac{100+2pt}{100}\right) \Rightarrow \left(\frac{100+2pt}{100}\right) = 2$ *[1 mark]*
$\Rightarrow 200 = 100 + 2pt \Rightarrow 2t = \frac{100}{p} \Rightarrow t = \frac{50}{p}$ *[1 mark]*
Substituting in $t = \log_{\left(\frac{100+q}{100}\right)} 2$ gives:
$\log_{\left(\frac{100+q}{100}\right)} 2 = \frac{50}{p} \Rightarrow 2 = \left(\frac{100+q}{100}\right)^{\frac{50}{p}}$ *[1 mark]*
$\Rightarrow 2^{\frac{p}{50}} = \frac{100+q}{100} \Rightarrow q = 100 \times 2^{\frac{p}{50}} - 100$
$\Rightarrow q = 100(2^{\frac{p}{50}} - 1)$ as required *[1 mark]*
[4 marks available in total — as above]

13 a) Use the quotient rule with:
$u = x^2\cos x \Rightarrow \frac{du}{dx} = (2x\cos x) + (x^2(-\sin x))$
$= 2x\cos x - x^2\sin x$ *[1 mark]*
$v = 3\sin x \Rightarrow \frac{dv}{dx} = 3\cos x$ *[1 mark]*
$\frac{dy}{dx} = \frac{v\frac{du}{dx} - u\frac{dv}{dx}}{v^2}$
$= \frac{(3\sin x)(2x\cos x - x^2\sin x) - (x^2\cos x)(3\cos x)}{(3\sin x)^2}$ *[1 mark]*
$= \frac{6x\sin x\cos x - 3x^2\sin^2 x - 3x^2\cos^2 x}{9\sin^2 x}$
$= \frac{x(2\sin x\cos x - x(\sin^2 x + \cos^2 x))}{3\sin^2 x}$ *[1 mark]*
Using $2\sin x\cos x = \sin 2x$ and $\sin^2 x + \cos^2 x = 1$:
$= \frac{x(\sin 2x - x)}{3\sin^2 x}$ as required. *[1 mark]*
[5 marks available in total — as above]
Alternatively, you could write $\frac{x^2\cos x}{3\sin x} = \frac{1}{3}x^2\cot x$ and use the product rule to differentiate.

b) At stationary points, $\frac{dy}{dx} = 0$. If $a = 0.95$ to 2 decimal places, then $0.945 \leq a < 0.955$. *[1 mark]*
When $x = 0.945$, $\frac{dy}{dx} = 0.00215...$ (positive)
When $x = 0.955$, $\frac{dy}{dx} = -0.00572...$ (negative)
There is a change of sign between 0.945 and 0.955, so $a = 0.95$ to 2 decimal places. *[1 mark]*
[2 marks available in total — as above]

c) When $x = 0.95$, $\frac{dy}{dx} = -0.00177...$ (negative) so the root is between 0.945 and 0.95, i.e. 0.95 is an overestimate. *[1 mark]*

14 a) $\frac{dV}{dt} = \frac{2}{Vt} \Rightarrow \int \frac{V}{2} dV = \int \frac{1}{t} dt$
$\Rightarrow \frac{V^2}{4} = \ln|t| + C = \ln t + C$ as $t > 0$
When $t = 1$, there are 4000 views so $V = 4$.
So $\frac{4^2}{4} = \ln 1 + C \Rightarrow C = 4 \Rightarrow \frac{V^2}{4} = \ln t + 4$
$\Rightarrow V^2 = 4(\ln t + 4) \Rightarrow V = 2\sqrt{\ln t + 4}$
At the end of day 7, $V = 2\sqrt{\ln 7 + 4} = 4.876...$
At the end of day 8, $V = 2\sqrt{\ln 8 + 4} = 4.931...$
So during day 8 it got $1000 \times (4.931... - 4.876...)$
$= 54.457... = 54$ views.
[5 marks available — 1 mark for the correct method for integration, 1 mark for the correct integration, 1 mark for finding the value of C, 1 mark for a correct method to find the number of views on day 8, 1 mark for the correct answer]

b) E.g. The model is undefined when $\ln t < -4$ (i.e. $t < 0.0183...$). Theo could improve the model by giving a different equation for V during this time (e.g. $V = 0$ for $t < 0.0183...$).
[2 marks available — 1 mark for a limitation of the model, 1 mark for a suitable improvement to address the given limitation]
There are other limitations that you could mention — for example, the model suggests that the views will continue to increase forever, so an upper limit on t might be needed.

15 a) $\cos 2x = \cos^2 x - \sin^2 x$
$= (1 - \sin^2 x) - \sin^2 x = 1 - 2\sin^2 x$ *[1 mark]*
$\Rightarrow \sin^2 x = \frac{1 - \cos 2x}{2}$
$\Rightarrow \sin^2 5x = \frac{1 - \cos 10x}{2} = \frac{1}{2}(1 - \cos 10x)$ *[1 mark]*
$\int \sin^2 5x \, dx = \frac{1}{2} \int 1 - \cos 10x \, dx$
$= \frac{1}{2}\left(x - \frac{\sin 10x}{10}\right) + C$ or $\frac{x}{2} - \frac{\sin 10x}{20} + C$
[1 mark for $\frac{x}{2}$, 1 mark for $-\frac{\sin 10x}{20}$]
[4 marks available in total — as above]

b) Using integration by parts: $u = x$, $\frac{dv}{dx} = \sin^2 5x$
$\frac{du}{dx} = 1$ and $v = \frac{x}{2} - \frac{\sin 10x}{20}$
$\int_0^{\frac{2\pi}{5}} x\sin^2 5x \, dx = \left[x\left(\frac{x}{2} - \frac{\sin 10x}{20}\right)\right]_0^{\frac{2\pi}{5}} - \int_0^{\frac{2\pi}{5}} 1\left(\frac{x}{2} - \frac{\sin 10x}{20}\right) dx$
$= \left[\frac{x^2}{2} - \frac{x\sin 10x}{20} - \left(\frac{x^2}{4} + \frac{\cos 10x}{200}\right)\right]_0^{\frac{2\pi}{5}}$
$= \left[\frac{x^2}{4} - \frac{x\sin 10x}{20} - \frac{\cos 10x}{200}\right]_0^{\frac{2\pi}{5}}$
$= \left[\frac{\left(\frac{2\pi}{5}\right)^2}{4} - \frac{\frac{2\pi}{5}\sin\frac{10 \times 2\pi}{5}}{20} - \frac{\cos\frac{10 \times 2\pi}{5}}{200}\right]$
$- \left[\frac{0^2}{4} - \frac{0\sin(10 \times 0)}{20} - \frac{\cos(10 \times 0)}{200}\right]$
$= \left[\frac{\pi^2}{25} - 0 - \frac{1}{200}\right] - \left[0 - 0 - \frac{1}{200}\right] = \frac{\pi^2}{25}$
[5 marks available — 1 mark for attempting to use integration by parts, 1 mark for applying the integration by parts formula correctly, 1 mark for the correct integral, 1 mark for substituting in limits of the integral, 1 mark for the correct answer]

Set 2 Paper 2 — Pure Mathematics 2

1 a) $\overrightarrow{AB} = \begin{pmatrix} -4 \\ 1 \\ 1 \end{pmatrix} - \begin{pmatrix} 4 \\ 1 \\ 5 \end{pmatrix} = \begin{pmatrix} -8 \\ 0 \\ -4 \end{pmatrix}$ *[1 mark]*
$|\overrightarrow{AB}| = \sqrt{(-8)^2 + 0^2 + (-4)^2}$ *[1 mark]* $= \sqrt{80} = 4\sqrt{5}$ *[1 mark]*
[3 marks available in total — as above]

b) $\overrightarrow{AC} = 3 \times \begin{pmatrix} -8 \\ 0 \\ -4 \end{pmatrix} = \begin{pmatrix} -24 \\ 0 \\ -12 \end{pmatrix}$ *[1 mark]*
$\overrightarrow{OC} = \begin{pmatrix} 4 \\ 1 \\ 5 \end{pmatrix} + \begin{pmatrix} -24 \\ 0 \\ -12 \end{pmatrix} = \begin{pmatrix} -20 \\ 1 \\ -7 \end{pmatrix}$ *[1 mark]*
[2 marks available in total — as above]

2 $64^a \times \left(\frac{1}{16}\right)^b \div \sqrt[c]{32} = (2^6)^a \times (2^{-4})^b \div (2^5)^{\frac{1}{c}}$
$= 2^{6a} \times 2^{-4b} \div 2^{\frac{5}{c}} = 2^{6a - 4b - \frac{5}{c}}$
So $d = 6a - 4b - \frac{5}{c}$
[3 marks available — 1 mark for correctly rewriting one term as a power of 2, 1 mark for expressing all terms as powers of 2, 1 mark for the correct answer]

3 a) £5069 is the amount of money that the farmer would make if she sold her maize crop when $t = 0$ (on 1st July). *[1 mark]*

b) $\frac{dP}{dt} = -2t + 66$ *[1 mark]* At stationary points, $\frac{dP}{dt} = 0$
$\Rightarrow 0 = -2t + 66 \Rightarrow t = 33$ *[1 mark]*
Since P is a quadratic with a negative coefficient of t^2, the turning point is a maximum *[1 mark]*, so the optimum selling date is 33 days after 1st July, which is 3rd August. *[1 mark]*
[4 marks available in total — as above]
You could also have justified your answer by finding the second derivative and showing it's negative at $t = 33$, so it's a maximum.

c) Substituting $t = 33$ into the equation for P gives $P = -(33^2) + (66 \times 33) + 5069 = £6158$. *[1 mark]*

d) For sufficiently large t, e.g. $t = 200$, P is negative which doesn't make sense as P is the amount she sells the crop for. The value of t could be restricted in order to improve the model, e.g. by making $0 \leq t \leq 111$ as $t = 111$ is the last day where P is positive.
[2 marks available — 1 mark for any suitable limitation, 1 mark for a sensible suggestion for how it can be improved]

4 $4\cos x - 11 = \frac{\sin^2 x - 3}{\cos x}$
$\Rightarrow 4\cos^2 x - 11\cos x = \sin^2 x - 3$ *[1 mark]*
$\Rightarrow 4\cos^2 x - 11\cos x = (1 - \cos^2 x) - 3$ *[1 mark]*
$\Rightarrow 5\cos^2 x - 11\cos x + 2 = 0$ *[1 mark]*
$\Rightarrow (5\cos x - 1)(\cos x - 2) = 0$ *[1 mark]*
$\Rightarrow \cos x = 0.2$ *[1 mark]* ($\cos x \neq 2$ as $-1 \leq \cos x \leq 1$)
$\Rightarrow x = \cos^{-1}(0.2) = 78.46...° = 78.5°$ (1 d.p.)
and $x = 360° - 78.46...° = 281.53...° = 281.5°$ (1 d.p.)
[1 mark for both]
[6 marks available in total — as above]

5 a) When $x = 0$, $y = |3(0) - 1| = 1$
When $y = 0$, $0 = |3x - 1| \Rightarrow 3x = 1 \Rightarrow x = \frac{1}{3}$

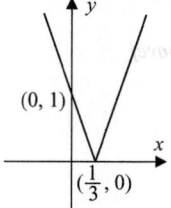

[3 marks available — 1 mark for v-shaped graph above the x-axis, 1 mark for a graph touching the x-axis at $x = \frac{1}{3}$, 1 mark for a graph crossing the y-axis at $y = 1$]
You could also start with the graph $y = 3x - 1$, which has gradient 3 and y-intercept -1. Then any point below the x-axis gets reflected in the x-axis to produce the graph of $y = |3x - 1|$.

b) First, solve the equation $|3x - 1| = 2x + 5$:
When $3x - 1 \geq 0$, the equation becomes
$3x - 1 = 2x + 5 \Rightarrow x = 6$ *[1 mark]*
When $3x - 1 < 0$, the equation becomes
$-3x + 1 = 2x + 5 \Rightarrow -5x = 4 \Rightarrow x = -0.8$ *[1 mark]*
Since it is given that the equation has 2 solutions, they must be at $x = 6$ and $x = -0.8$
As the graph of $y = |3x - 1|$ is v-shaped, the inequality $|3x - 1| \leq 2x + 5$ will be satisfied between the two solutions, i.e. when $-0.8 \leq x \leq 6$ *[1 mark]*
[3 marks available in total — as above]

6 $S_{20} = 1390$ and $S_{30} = 3135$
Using the formula $S_n = \frac{1}{2}n[2a + (n - 1)d]$:
(1): For S_{20}, $1390 = 10(2a + 19d) = 20a + 190d$ *[1 mark]*
(2): For S_{30}, $3135 = 15(2a + 29d) = 30a + 435d$ *[1 mark]*
(1) × 3: $4170 = 60a + 570d$
(2) × 2: $6270 = 60a + 870d$
Now subtract one equation from the other to give:
$2100 = 300d \Rightarrow d = 7$ *[1 mark]* Substituting back into (1) gives:
$1390 = 20a + 190 \times 7 \Rightarrow 60 = 20a \Rightarrow a = 3$ *[1 mark]*
Now if $a = 3$ and $d = 7$, $u_{10} = 3 + (9 \times 7) = 66$ prizes *[1 mark]*
[5 marks available in total — as above]

7 a) Rearranging $x = 3t + 1$ to make t the subject gives $\frac{x-1}{3} = t$.
Substituting this into $y = (t + 3)^3 - 5$ gives
$y = \left(\frac{x-1}{3} + 3\right)^3 - 5 = \left(\frac{x-1}{3} + \frac{9}{3}\right)^3 - 5 = \left(\frac{x+8}{3}\right)^3 - 5$
[2 marks available — 1 mark for rearranging to make t the subject, 1 mark for substituting and rearranging to give the required result]

b) Caleb hasn't changed the limits of the integration — he needs to change them to be in terms of t *[1 mark]*.
He also hasn't multiplied by $\frac{dx}{dt}$ *[1 mark]*.
[2 marks available in total — as above]

c) The limits of the integration become:
$x = 4 \Rightarrow 4 = 3t + 1 \Rightarrow t = 1$
$x = 10 \Rightarrow 10 = 3t + 1 \Rightarrow t = 3$
$\frac{dx}{dt} = 3$ and $y = (t + 3)^3 - 5$
So $\int_4^{10} y \, dx = 3\int_1^3 ((t+3)^3 - 5) \, dt = 3\left[\frac{1}{4}(t+3)^4 - 5t\right]_1^3$
$= 3\left[\frac{6^4}{4} - 15 - \left(\frac{4^4}{4} - 5\right)\right] = 750$
[3 marks available — 1 mark for correcting both of Caleb's errors (i.e. converting the limits and finding $\frac{dx}{dt}$), 1 mark for integrating correctly, 1 mark for substituting in the limits to obtain the correct final answer]

8 a) $y = at^b \Rightarrow \log_{10} y = \log_{10}(at^b) \Rightarrow \log_{10} y = \log_{10} a + \log_{10} t^b$
$\Rightarrow \log_{10} y = \log_{10} a + b \log_{10} t$
[2 marks available — 1 mark for taking logs of both sides, 1 mark for using laws of logs to simplify]

b) Use the graph to find the values of $\log_{10} a$ and b:
$\log_{10} a$ is the vertical intercept, which is 2.475 *[1 mark]*.
b = gradient = $\frac{\text{change in } \log_{10} y}{\text{change in } \log_{10} t} = \frac{2.60 - 2.55}{0.25 - 0.15} = 0.5$ *[1 mark]*
So the equation of the line of best fit is:
$\log_{10} y = 2.475 + 0.5 \log_{10} t$
When $y = 1000$: $\log_{10} 1000 = 2.475 + 0.5 \log_{10} t$ *[1 mark]*
$\Rightarrow 0.5 \log_{10} t = 3 - 2.475 \Rightarrow \log_{10} t = 1.05$
$\Rightarrow t = 11.220... = 11$ days (nearest whole day) *[1 mark]*
[4 marks available in total — as above]
You could also work out the value of a and use the original equation ($y = 298.53... \times t^{0.5}$) to find the value of t.

c) E.g. The observed pattern might not continue — the rate of change could increase if the ants breed faster or more join the colony, or decrease if the breeding rate or number of ants joining the colony slows. / The colony might reach a certain size then remain at that size (e.g. due to restrictions on space or resources). / Once it gets to a certain size, ants might leave the colony to form a new one, so the number of ants could decrease. *[1 mark for a sensible limitation linked to the number of ants in the colony]*

9 a) The graph of $y = \frac{-1}{x + 5} + 2$ will be a translation of $y = \frac{-1}{x}$ 5 units to the left and 2 units up. So the equations of the asymptotes will be $x = -5$ and $y = 2$.

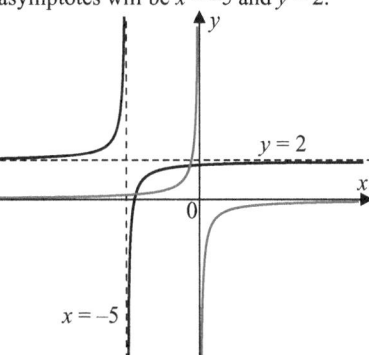

[2 marks available — 1 mark for drawing the graph in the correct position, 1 mark for drawing and labelling the asymptotes correctly]

b) A stretch, scale factor 3, in the y-direction *[1 mark]*
followed by a translation by the vector $\binom{0}{2}$ *[1 mark]*.
[2 marks available in total — as above]
You could also have described it as a translation by the vector $\binom{0}{\frac{2}{3}}$ followed by a stretch in the y-direction by scale factor 3.

10 a) The curve has a root at -1 so $(x + 1)$ is a factor. *[1 mark]*
$x^3 - 5x^2 + 2x + 8 = (x + 1)(x^2 - 6x + 8)$ *[1 mark]*
$= (x + 1)(x - 4)(x - 2)$
The x-value at point C must be $x = 2$,
so the coordinates of C are (2, 0). *[1 mark]*
[3 marks available in total — as above]

b) Shaded area = $\left(\frac{1}{2} \times 1 \times 6\right) + \int_1^2 (x^3 - 5x^2 + 2x + 8) \, dx$
[1 mark]
$= 3 + \left[\frac{x^4}{4} - \frac{5x^3}{3} + x^2 + 8x\right]_1^2$ *[1 mark]*
$= 3 + \left[\frac{16}{4} - \frac{40}{3} + 4 + 16\right] - \left[\frac{1}{4} - \frac{5}{3} + 1 + 8\right]$ *[1 mark]*
$= 3 + \frac{32}{3} - \frac{91}{12} = \frac{73}{12}$ *[1 mark]*
[4 marks available in total — as above]

11 a) Let $f(x) = x^3$ then
$f'(x) = \lim_{h \to 0}\left(\frac{(x+h)^3 - x^3}{h}\right)$ *[1 mark]*
$= \lim_{h \to 0}\left(\frac{(x^2 + 2xh + h^2)(x+h) - x^3}{h}\right)$
$= \lim_{h \to 0}\left(\frac{(x^3 + 3x^2h + 3xh^2 + h^3) - x^3}{h}\right)$ *[1 mark]*
$= \lim_{h \to 0}\left(\frac{3x^2h + 3xh^2 + h^3}{h}\right)$
$= \lim_{h \to 0}(3x^2 + 3xh + h^2)$ *[1 mark]*
As $h \to 0$, $3x^2 + 3xh + h^2 \to 3x^2$, so $f'(x) = 3x^2$ *[1 mark]*
[4 marks available in total — as above]

b) Let $u = \sin x \Rightarrow \frac{du}{dx} = \cos x$
$y = \sin^3 x = u^3$ so $\frac{dy}{du} = 3u^2 = 3\sin^2 x$
Then $\frac{dy}{dx} = \frac{dy}{du} \times \frac{du}{dx} = 3\cos x \sin^2 x$
Use the product rule where $u = 3\cos x$ and $v = \sin^2 x$:
$\Rightarrow \frac{du}{dx} = -3\sin x$ and $\frac{dv}{dx} = 2\sin x \cos x$

$\frac{d^2y}{dx^2} = v\frac{du}{dx} + u\frac{dv}{dx} = \sin^2 x \times -3\sin x + 3\cos x \times 2\sin x\cos x$
$= -3\sin^3 x + 6\sin x\cos^2 x$

When $x = \frac{\pi}{2}$, $\frac{d^2y}{dx^2} = -3\sin^3\left(\frac{\pi}{2}\right) + 6\sin\left(\frac{\pi}{2}\right)\cos^2\left(\frac{\pi}{2}\right)$
$= -3(1) + 6(0) = -3$

So the stationary point is a maximum.
[5 marks available — 1 mark for using a suitable method to find the first derivative, 1 mark for finding the first derivative correctly, 1 mark for using a suitable method to find the second derivative, 1 mark for finding the second derivative correctly, 1 mark for the correct answer and conclusion]
There are other methods you could have used to differentiate, such as using trig identities on the expressions first. You might get a $\frac{d^2y}{dx^2}$ that looks different to the one above, but it should still give you a value of −3 at the stationary point.

12 a) Area of minor sector $AOB = \frac{1}{2}r^2\theta$
Area of triangle $AOB = \frac{1}{2}r^2\sin\theta$
Area of $S_1 = \frac{1}{2}r^2\theta - \frac{1}{2}r^2\sin\theta$ *[1 mark]*
Area of whole circle $= \pi r^2$
Now $S_1 : S_2 = 2 : 7 \Rightarrow S_1 :$ whole circle $= 2 : 9$
$\Rightarrow S_1 = \frac{2}{9} \times$ area of whole circle *[1 mark]*
$\Rightarrow \frac{1}{2}r^2\theta - \frac{1}{2}r^2\sin\theta = \frac{2}{9} \times \pi r^2$ *[1 mark]*
$\Rightarrow \frac{1}{2}\theta - \frac{1}{2}\sin\theta = \frac{2}{9}\pi$
$\Rightarrow \theta - \sin\theta = \frac{4\pi}{9}$
$\Rightarrow \theta - \sin\theta - \frac{4\pi}{9} = 0$ as required *[1 mark]*
[4 marks available in total — as above]
You could also have found an expression for the area of S_2 and used the ratio to relate it to S_1 or the area of the whole circle.

b) $\theta_0 = \frac{\pi}{2} \Rightarrow \theta_1 = \sin\frac{\pi}{2} + \frac{4\pi}{9} = 2.3962...$
$\theta_2 = 2.0744...$ $\theta_3 = 2.2720...$ $\theta_4 = 2.1602...$ $\theta_5 = 2.2274...$
So $\theta = 2.2$ radians (2 s.f.) because θ_4 and θ_5 round to the same number to 2 s.f.
[2 marks available — 1 mark for using the iterative formula correctly, 1 mark for the correct answer with justification]

13 a) $\frac{-x-8}{x^2+6x+8} = \frac{-x-8}{(x+4)(x+2)} \equiv \frac{C}{x+4} + \frac{D}{x+2}$ *[1 mark]*
$\Rightarrow -x - 8 \equiv C(x+2) + D(x+4)$
When $x = -4$: $4 - 8 = C(-4+2) \Rightarrow -4 = -2C \Rightarrow C = 2$
When $x = -2$: $2 - 8 = D(-2+4) \Rightarrow -6 = 2D \Rightarrow D = -3$
[1 mark for the correct value of C or D]
$\frac{-x-8}{(x+4)(x+2)} = \frac{2}{x+4} - \frac{3}{x+2}$ *[1 mark]*
[3 marks available in total — as above]

b) $\frac{2}{x+4} - \frac{3}{x+2} = 2(x+4)^{-1} - 3(x+2)^{-1}$
$= 2(4^{-1})\left(\frac{x}{4}+1\right)^{-1} - 3(2^{-1})\left(\frac{x}{2}+1\right)^{-1}$ *[1 mark]*

$2(4^{-1})\left(\frac{x}{4}+1\right)^{-1} = \frac{1}{2}\left(1 - \frac{x}{4} + \frac{-1 \times -2}{1 \times 2}\left(\frac{x}{4}\right)^2 + ...\right)$
$= \frac{1}{2} - \frac{x}{8} + \frac{x^2}{32} + ...$ *[1 mark]*

$3(2^{-1})\left(\frac{x}{2}+1\right)^{-1} = \frac{3}{2}\left(1 - \frac{x}{2} + \frac{-1 \times -2}{1 \times 2}\left(\frac{x}{2}\right)^2 + ...\right)$
$= \frac{3}{2} - \frac{3x}{4} + \frac{3x^2}{8} + ...$ *[1 mark]*

So $2(x+4)^{-1} - 3(x+2)^{-1} \approx \frac{1}{2} - \frac{x}{8} + \frac{x^2}{32} - \left(\frac{3}{2} - \frac{3x}{4} + \frac{3x^2}{8}\right)$
$= -1 + \frac{5x}{8} - \frac{11x^2}{32}$ *[1 mark]*
[4 marks available in total — as above]

c) The expansion of $\left(\frac{x}{4}+1\right)^{-1}$ is valid when:
$\left|\frac{x}{4}\right| < 1 \Rightarrow -1 < \frac{x}{4} < 1 \Rightarrow -4 < x < 4$ *[1 mark]*
The expansion of $\left(\frac{x}{2}+1\right)^{-1}$ is valid when:
$\left|\frac{x}{2}\right| < 1 \Rightarrow -1 < \frac{x}{2} < 1 \Rightarrow -2 < x < 2$ *[1 mark]*

So the expansion of f(x) is valid when both of these are satisfied, i.e. when $-2 < x < 2$. In set notation, this is:
$\{x : -2 < x < 2\}$ or $\{x : x > -2\} \cap \{x : x < 2\}$ *[1 mark]*
[3 marks available in total — as above]

d) When $x = 0.1$, $\frac{-0.1-8}{0.1^2+6(0.1)+8} = -0.9407...$
and $-1 + \frac{5(0.1)}{8} - \frac{11(0.1)^2}{32} = -0.9409...$ *[1 mark for both]*
Percentage error $= \frac{-0.9409... - -0.9407...}{-0.9407...} \times 100\%$
$= 0.018\%$ (2 s.f.) *[1 mark]*
[2 marks available in total — as above]

14 a) E.g. Using $\tan^2\theta + 1 = \sec^2\theta$ and $\cos 2\theta = \cos^2\theta - \sin^2\theta$
$\frac{1-\tan^2\theta}{1+\tan^2\theta} = \frac{1-\tan^2\theta}{\sec^2\theta}$ *[1 mark]* $= \frac{1}{\sec^2\theta} - \frac{\tan^2\theta}{\sec^2\theta}$ *[1 mark]*
$= \cos^2\theta - \frac{\sin^2\theta}{\cos^2\theta} \div \frac{1}{\cos^2\theta} = \cos^2\theta - \frac{\sin^2\theta}{\cos^2\theta} \times \cos^2\theta$ *[1 mark]*
$= \cos^2\theta - \sin^2\theta$ *[1 mark]* $= \cos 2\theta$
[4 marks available in total — as above]

b) From part a), $\frac{1-\tan^2\left(\beta+\frac{\pi}{2}\right)}{1+\tan^2\left(\beta+\frac{\pi}{2}\right)} = \cos 2\left(\beta+\frac{\pi}{2}\right)$
$\Rightarrow \cos(2\beta+\pi) - 0.5\sec(2\beta+\pi) = 0$ *[1 mark]*
$\Rightarrow \cos(2\beta+\pi) - \frac{0.5}{\cos(2\beta+\pi)} = 0$
$\Rightarrow \cos^2(2\beta+\pi) - 0.5 = 0$ *[1 mark]*
$\Rightarrow \cos(2\beta+\pi) = \pm\sqrt{0.5}$ *[1 mark]*
$\Rightarrow 2\beta + \pi = \cos^{-1}(\pm\sqrt{0.5})$ and $0 < \beta < \pi \Rightarrow \pi < 2\beta + \pi < 3\pi$
$\Rightarrow 2\beta + \pi = \frac{5\pi}{4}, \frac{7\pi}{4}, \frac{9\pi}{4}, \frac{11\pi}{4}$ *[1 mark]*
$\Rightarrow 2\beta = \frac{\pi}{4}, \frac{3\pi}{4}, \frac{5\pi}{4}, \frac{7\pi}{4} \Rightarrow \beta = \frac{\pi}{8}, \frac{3\pi}{8}, \frac{5\pi}{8}$ and $\frac{7\pi}{8}$ *[1 mark]*
[5 marks available in total — as above]

15 a) $\frac{dC}{dt} = -kCt \Rightarrow \int \frac{1}{C}\,dC = \int -kt\,dt$
$\Rightarrow \ln|C| = -\frac{k}{2}t^2 + d$ *[1 mark]*
$\Rightarrow C = e^{-\frac{k}{2}t^2+d} = Ae^{-\frac{k}{2}t^2}$ (where $A = e^d$) *[1 mark]*
At the end of the holidays $t = 0$, so $A = 3600$.
The equation for C is $C = 3600e^{-\frac{k}{2}t^2}$ *[1 mark]*.
[3 marks available in total — as above]

b) $k = 0.2$, so the equation for C becomes $C = 3600e^{-0.1t^2}$.
Solve the equation $300 = 3600e^{-0.1t^2}$:
$300 = 3600e^{-0.1t^2} \Rightarrow \frac{1}{12} = e^{-0.1t^2}$
$\Rightarrow \ln\frac{1}{12} = -0.1t^2 \Rightarrow t = \sqrt{-10\ln\frac{1}{12}} = 4.984...$
So $t = 5$ weeks (to the nearest whole week).
[3 marks available — 1 mark for substituting the values of k and C into the equation, 1 mark for rearranging and taking logs, 1 mark for the correct answer]
You can check that your answer is correct by putting t = 4 and t = 5 into the equation for C: t = 4 gives 726.827... customers, and t = 5 gives 295.505... This is less than 300, so the ice cream parlour will close after 5 weeks.

Set 2 Paper 3 — Statistics and Mechanics

1 a) Since values are given correct to the nearest knot, speeds in this class are actually between 4.5 kn and 8.5 kn, giving a class width of 4 kn. *[1 mark]*

b) $n = 17 + 14 + 7 + 2 = 40$, so Q_1 is in the $\frac{40}{4} = 10^{\text{th}}$ position and Q_3 is in the $\frac{40 \times 3}{4} = 30^{\text{th}}$ position. From the frequency table, Q_1 is in the 0 – 4 class and Q_3 is in the 5 – 8 class.
[1 mark for both]
Using linear interpolation:
$Q_1 \approx 0 + 4.5 \times \frac{10-0}{17} = 2.6470... = 2.65$ kn (3 s.f.) *[1 mark]*
$Q_3 \approx 4.5 + 4 \times \frac{30-17}{14} = 4.5 + 3.7142... = 8.2142...$
$= 8.21$ kn (3 s.f.) *[1 mark]*

IQR ≈ 8.2142... – 2.6470... = 5.5672...
= 5.57 kn (3 s.f.) *[1 mark]*
[4 marks available in total — as above]

c) Lower boundary = 2.6470... – (1.5 × 5.5672...) = –5.7037...
The lowest data value, 1, is greater than the lower boundary, so it is not an outlier. *[1 mark]*
Upper boundary = 8.2142... + (1.5 × 5.5672...) = 16.565...
The highest data value, 18, is greater than the upper boundary, so it is an outlier. *[1 mark]*
[2 marks available in total — as above]

d) E.g. The median, as it is not as heavily affected by outliers. *[1 mark]*
You could also argue that the mean is better because it uses all of the values so may be more representative of the data — as long as you justify your answer, you'll get the mark.

e) E.g. He has only looked at data for part of each country — the average for the whole country might be different / he has only looked at data for part of 1987 — the weather patterns may have changed since then / he has only looked at windspeed — other data may contradict this / he has not done a hypothesis test to see whether the higher mean is just down to random chance / he has not considered any other measures of location or variation (e.g. median, interquartile range, etc) / he has not considered other factors that could impact windspeed (e.g. how close the location is to the coast) / etc.
[2 marks available — 1 mark for each sensible criticism]

2 a) Draw a Venn diagram with three overlapping circles, then fill in the values of n(only R), n(don't like any), n(only R and T), n(only S and T) and n(only S and R):

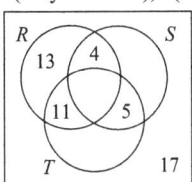 *[1 mark]*

n(R and S) = 26, so n(R and S and T) = 26 – 4 = 22 *[1 mark]*
n(S) = 40, so n(only S) = 40 – (4 + 5 + 22) = 9 *[1 mark]*
100 students were surveyed, so n(only T)
= 100 – (13 + 4 + 9 + 11 + 22 + 5 + 17) = 19 *[1 mark]*
So the completed Venn diagram is:

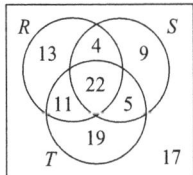

[4 marks available in total — as above]

b) i) P(likes at least one) = 1 – P(doesn't like any)
= $1 - \frac{17}{100} = \frac{83}{100}$
[2 marks available — 1 mark for the correct numerator, 1 mark for the correct denominator]
You could also add up all of the numbers inside the circles (you should get 83) and divide it by the total (100).

ii) $P(T | S) = \frac{5 + 22}{9 + 5 + 4 + 22} = \frac{27}{40}$
[2 marks available — 1 mark for the correct numerator, 1 mark for the correct denominator]

iii) $P((R \cap S') | T) = \frac{11}{19 + 11 + 22 + 5} = \frac{11}{57}$
[2 marks available — 1 mark for the correct numerator, 1 mark for the correct denominator]

c) If the teacher's claim is true, then the events T and S are independent, i.e. $P(T) \times P(S) = P(T \cap S)$.
$P(T) \times P(S) = \frac{11 + 22 + 5 + 19}{100} \times \frac{4 + 22 + 5 + 9}{100} = \frac{57}{250}$

$P(T \cap S) = \frac{27}{100}$ so the results of the survey do not support the teachers claim.
[3 marks available — 1 mark for using appropriate probability law for determining whether events are dependent, 1 mark for correctly calculating relevant probabilities, 1 mark for correct conclusion]
There are other ways you could check whether these are independent — for example, you could compare P(T|S) to P(T).

3 a) E.g. The graph shows weak positive correlation between the variables. *[1 mark]*

b) $H_0: \rho = 0$, $H_1: \rho \neq 0$ *[1 mark]*
This is a two-tailed test at the 10% significance level, so use the 0.05 column in the critical value table.
$n = 10$, so critical value = 0.5494. *[1 mark]*
0.5494 > 0.3850, so there is not enough evidence at the 10% level to reject H_0 and conclude that the product moment correlation coefficient is non-zero. *[1 mark]*
[3 marks available in total — as above]

c) The remaining points would lie closer to a straight line, so the value of r would increase. *[1 mark]*

4 a) Any one of: there are a fixed number of trials (packets of sweets) / each trial is independent of the others (whether one packet contains a gold-wrapped sweet does not affect whether another packet does) / there are only two possible outcomes (a packet either contains a gold-wrapped sweet or doesn't) / the probability remains constant from trial to trial. *[1 mark]*
$n = 50$, $p = \frac{9}{20} = 0.45$ *[1 mark for both]*
[2 marks available in total — as above]

b) $H_0: p = 0.45$, $H_1: p < 0.45$ *[1 mark]* $G \sim B(50, 0.45)$
Using the binomial tables: $P(G \leq 16) = 0.0427$ *[1 mark]*
0.0427 < 0.05, so the result is significant. *[1 mark]*
There is enough evidence at the 5% level to reject H_0 and conclude that the company has put fewer gold-wrapped sweets in their packets of sweets than they said they did. *[1 mark]*
[4 marks available in total — as above]

5 a) i) P(positive test result and carrier)
= 0.98 × 0.03 = 0.0294 *[1 mark]*
P(positive test result and non-carrier)
= (1 – 0.95) × (1 – 0.03) = 0.0485 *[1 mark]*
P(positive test result) = 0.0294 + 0.0485 = 0.0779 *[1 mark]*
P(carrier given a positive test result)
= 0.0294 ÷ 0.0779 = 0.37740... = 0.377 (3 s.f.) *[1 mark]*
[4 marks available in total — as above]

ii) E.g. Joey is a random member of the UK population and isn't more likely to have the disease just because he takes the test. *[1 mark]*

b) P(correct test result) = P(positive test result and carrier) + P(negative test result and not carrier) *[1 mark]*
= (0.98 × 0.03) + (0.95 × 0.97) = 0.0294 + 0.9215
= 0.9509 or 95.09% so Hiroshi is not correct. *[1 mark]*
[2 marks available in total — as above]

c) E.g. No, because the majority of positive results are for people who don't carry the disease (i.e. most of the time a positive result is a false positive). *[1 mark]*
You could argue either yes or no for this question, as long as you fully explain your reasoning. E.g. you could say yes because only 2% of people who carry the disease will not be identified.

6 a) Let L be the length of a pencil in cm. Then $L \sim N(18, 0.1^2)$.
P(pencil will fit in box) = $P(L < 18.2)$ *[1 mark]*
From your calculator: $P(L < 18.2) = 0.9772...$ *[1 mark]*
P(box will not be damaged) = $(0.9772...)^5 = 0.8913...$ *[1 mark]*
P(box will be damaged) = 1 – 0.8913...
= 0.1086... = 0.109 (3 s.f.) *[1 mark]*
[4 marks available in total — as above]
You could also have used a binomial distribution for the number of pencils in a box that are longer/shorter than 18.2 cm.

b) Let X be the number of boxes that are damaged in one day.
Then $X \sim B(1000, 0.1086...)$ *[1 mark]*
$\Rightarrow \mu = 1000 \times 0.1086... = 108.69...$
and $\sigma^2 = 1000 \times 0.1086... \times (1 - 0.1086...) = 96.87...$
So X is approximated by $Y \sim N(108.69..., 96.87...)$ *[1 mark]*
$P(X > 125) \approx P(Y > 125.5)$ *[1 mark]*
From your calculator:
$P(Y > 125.5) = 0.043842... = 0.0438$ (3 s.f.) *[1 mark]*
[4 marks available in total — as above]

7 a) Resolving forces vertically: $R = W + \frac{5}{8}W = \frac{13}{8}W$
[2 marks available — 1 mark for a correct method to resolve forces vertically, 1 mark for the correct answer]

b) Let x be the distance from A to the centre of mass of the plank.
E.g. equating moments about the centre of the plank:
Clockwise moments = anticlockwise moments
$\frac{5}{8}W \times 1 = W \times (2 - x)$
$\Rightarrow \frac{5}{8} = 2 - x \Rightarrow x = 2 - \frac{5}{8} = 1\frac{3}{8}$ m = 1.375 m
[3 marks available — 1 mark for equating moments, 1 mark for a correct equation, 1 mark for the correct answer]

8 a) $\mathbf{s} = \mathbf{s}$, $\mathbf{u} = (1.5\mathbf{i} - 2\mathbf{j})$ ms^{-1}, $\mathbf{a} = 0$ ms^{-2}, $t = 7$ seconds
$\mathbf{s} = \mathbf{u}t + \frac{1}{2}\mathbf{a}t^2 = 7(1.5\mathbf{i} - 2\mathbf{j}) + 0 = (10.5\mathbf{i} - 14\mathbf{j})$ m
[2 marks available — 1 mark for a correct method, 1 mark for the correct answer]

b) i) They collide after $t = 7 + 6 = 13$ seconds, so the position of the skateboarder (and where they collide) is:
$\mathbf{s} = 13(1.5\mathbf{i} - 2\mathbf{j}) + 0 = (19.5\mathbf{i} - 26\mathbf{j})$
[2 marks available — 1 mark for a correct method, 1 mark for the correct answer]

ii) For the cyclist's motion between $t = 7$ s and $t = 13$ s:
$\mathbf{s} = \mathbf{s}$, $\mathbf{u} = (-4.5\mathbf{i} + 6\mathbf{j})$ ms^{-1}, $\mathbf{a} = (-0.3\mathbf{i} + 0.4\mathbf{j})$ ms^{-2},
$t = (13 - 7) = 6$ s. Now, using $\mathbf{s} = \mathbf{u}t + \frac{1}{2}\mathbf{a}t^2$
$\mathbf{s} = (-4.5\mathbf{i} + 6\mathbf{j}) \times 6 + \frac{1}{2} \times (-0.3\mathbf{i} + 0.4\mathbf{j}) \times 6^2$ *[1 mark]*
$= (-27\mathbf{i} + 36\mathbf{j}) + (-5.4\mathbf{i} + 7.2\mathbf{j})$
$= (-32.4\mathbf{i} + 43.2\mathbf{j})$ m *[1 mark]*
So the initial position of the cyclist is:
$(19.5\mathbf{i} - 26\mathbf{j}) - (-32.4\mathbf{i} + 43.2\mathbf{j})$
$= (51.9\mathbf{i} - 69.2\mathbf{j})$ m *[1 mark]*
[3 marks available in total — as above]

9 a) Resolving forces vertically on Q: $4g - T = 4a$ (1) *[1 mark]*
The force pulling P downwards parallel to the plane is $2g \sin 15°$. *[1 mark]* So resolving forces parallel to the plane on P: $T - 2g \sin 15° = 2a$ (2) *[1 mark]*
Adding equations (1) and (2) gives:
$4g - T + T - 2g \sin 15° = 4a + 2a$
$4g - 2g \sin 15° = 6a$ *[1 mark]*
$a = 34.127... \div 6 = 5.6878... = 5.7$ ms^{-2} (2 s.f.) *[1 mark]*
[5 marks available in total — as above]
You could have used $\cos 75°$ instead of $\sin 15°$.

b) Substituting $a = 5.6878...$ ms^{-2} into e.g. equation (2) gives:
$T = 2g \sin 15° + 2(5.6878...) = 5.0728... + 11.375...$
$= 16.448...$ N *[1 mark]*
Vertical component of $F = T + T \sin 15°$ *[1 mark]*
$= 16.448... + 16.448... \sin 15°$
$= 20.705... = 21$ N (2 s.f.) *[1 mark]*
[3 marks available in total — as above]

10 a) $\mathbf{r} = \int \mathbf{v}\,dt$ *[1 mark]* $= \int \begin{pmatrix} 5t - 6t^2 \\ 6 - t^3 \end{pmatrix} dt = \begin{pmatrix} \frac{5t^2}{2} - 2t^3 \\ 6t - \frac{t^4}{4} \end{pmatrix} + \mathbf{C}$ *[1 mark]*

At $t = 0$, $\mathbf{r} = \begin{pmatrix} \frac{5(0)^2}{2} - 2(0)^3 \\ 6(0) - \frac{(0)^4}{4} \end{pmatrix} + \mathbf{C} = \begin{pmatrix} 6 \\ 2 \end{pmatrix} \Rightarrow \mathbf{C} = \begin{pmatrix} 6 \\ 2 \end{pmatrix}$ *[1 mark]*

At $t = 3$, $\mathbf{r} = \begin{pmatrix} \frac{5(3)^2}{2} - 2(3)^3 + 6 \\ 6(3) - \frac{(3)^4}{4} + 2 \end{pmatrix} = \begin{pmatrix} -25.5 \\ -0.25 \end{pmatrix}$ m *[1 mark]*

[4 marks available in total — as above]

b) $\mathbf{a} = \frac{d\mathbf{v}}{dt} = \begin{pmatrix} 5 - 12t \\ -3t^2 \end{pmatrix}$ *[1 mark]*
At $t = 1$, $\mathbf{a} = \begin{pmatrix} 5 - 12(1) \\ -3(1)^2 \end{pmatrix} = \begin{pmatrix} -7 \\ -3 \end{pmatrix}$ ms^{-2} *[1 mark]*
$|\mathbf{a}| = \sqrt{(-7)^2 + (-3)^2}$ *[1 mark]*
$= \sqrt{49 + 9} = \sqrt{58} = 7.62$ ms^{-2} (3 s.f.) *[1 mark]*
[4 marks available in total — as above]

c) Using $\mathbf{F}_{net} = m\mathbf{a}$: $\begin{pmatrix} -20 \\ k \end{pmatrix} = 5 \begin{pmatrix} 5 - 12t \\ -3t^2 \end{pmatrix}$ *[1 mark]*
$-20 = 5(5 - 12t) \Rightarrow -4 = 5 - 12t$
$\Rightarrow 12t = 9 \Rightarrow t = 0.75$ seconds *[1 mark]*
So $k = 5(-3t^2) = -15(0.75)^2 = -15 \times 0.5625 = -8.4375$ *[1 mark]*
[3 marks available in total — as above]

11 a) Considering the vertical motion of the particle:
$s = -3.5$ m, $u_V = (203 \sin \theta)$ ms^{-1}, $a = -g = -9.8$ ms^{-2},
$t = 5$ seconds. Using $s = ut + \frac{1}{2}at^2$:
$-3.5 = (203 \sin \theta \times 5) + \left(\frac{1}{2} \times (-9.8) \times 5^2\right)$ *[1 mark]*
$\Rightarrow -3.5 = 1015 \sin \theta - 122.5 \Rightarrow 119 = 1015 \sin \theta$
$\Rightarrow \sin \theta = \frac{119}{1015}$ *[1 mark]*
$\Rightarrow \theta = \sin^{-1}(0.1172...) = 6.732...° = 6.73°$ (3 s.f.) *[1 mark]*
[3 marks available in total — as above]
Upward was defined as positive here — if you'd taken downwards to be positive, then s and a would have been positive, and u would have been negative (you'd still get the same answer).

b) There is no horizontal acceleration, so:
$v_H = u_H = 203 \cos \theta = 201.6$ ms^{-1} *[1 mark]*
Considering the vertical motion of the particle:
$s = -3.5$ m, $u_V = 203 \sin \theta = 23.8$ ms^{-1}, $v_V = v_V$,
$a = -g = -9.8$ ms^{-2}, $t = 5$ seconds.
Using $s = vt - \frac{1}{2}at^2$: $-3.5 = 5v_V + 122.5$ *[1 mark]*
$\Rightarrow 5v_V = -126 \Rightarrow v_V = -25.2$ ms^{-1} *[1 mark]*
Let α be the angle between the horizontal and the direction of the bullet's motion:
$\tan \alpha = \frac{v_V}{v_H} = \frac{25.2}{201.6}$
$\Rightarrow \alpha = \tan^{-1}(0.125) = 7.125...° = 7.13°$ (3 s.f.) *[1 mark]*
[4 marks available in total — as above]

c) v_H is constant, so minimum speed is when $v_V = 0$.
Using $v = u + at$ for the vertical motion:
$0 = 203 \sin \theta - 9.8t$ *[1 mark]*
$\Rightarrow 9.8t = 23.8 \Rightarrow t = 2.428... = 2.43$ s (3 s.f.) *[1 mark]*
[2 marks available in total — as above]

12 a) Resolving forces perpendicular to the plane:
$R = 2g \cos 20°$ *[1 mark]*
Resolving forces parallel to the plane: $F_{net} = ma$ *[1 mark]*
$2g \sin 20° - F = 2 \times 0.15g$ *[1 mark]*
$\Rightarrow F = 2g \sin 20° - 0.3g$
The particle is sliding, so $F = \mu R = \mu(2g \cos 20°)$ *[1 mark]*
$\Rightarrow 2g \sin 20° - 0.3g = \mu(2g \cos 20°)$ *[1 mark]*
$\mu = \frac{2 \sin 20° - 0.3}{2 \cos 20°} = 0.20434... = 0.20$ (2 d.p.) *[1 mark]*
[6 marks available in total — as above]

b) Resolving parallel to the plane: $F = 2g \sin \theta$ *[1 mark]*
Resolving perpendicular to the plane: $R = 2g \cos \theta$
Since the particle is not sliding, $F \leq \mu R$:
$2g \sin \theta \leq \mu(2g \cos \theta)$ *[1 mark]* $\Rightarrow \frac{2g \sin \theta}{2g \cos \theta} \leq 0.20434...$
$\Rightarrow \tan \theta \leq 0.20434...$ *[1 mark]* $\Rightarrow \theta \leq 11.549...°$ *[1 mark]*
[4 marks available in total — as above]

CGP

A-Level
Mathematics

Exam Board: Edexcel

Practice Exam Papers
Formula Booklet

Mathematical Formulae

Pure Mathematics

Mensuration

Surface area of sphere = $4\pi r^2$ Area of curved surface of cone = $\pi r \times$ slant height

Arithmetic Series

$S_n = \frac{1}{2}n(a + l) = \frac{1}{2}n[2a + (n-1)d]$

Geometric Series

$S_n = \frac{a(1 - r^n)}{1 - r}$ $S_\infty = \frac{a}{1 - r}$ for $|r| < 1$

Binomial Series

$(a + b)^n = a^n + \binom{n}{1}a^{n-1}b + \binom{n}{2}a^{n-2}b^2 + \ldots + \binom{n}{r}a^{n-r}b^r + \ldots + b^n$ $(n \in \mathbb{N})$, where $\binom{n}{r} = {}^nC_r = \frac{n!}{r!(n-r)!}$

$(1 + x)^n = 1 + nx + \frac{n(n-1)}{1 \times 2}x^2 + \ldots + \frac{n(n-1)\ldots(n-r+1)}{1 \times 2 \times \ldots \times r}x^r + \ldots$ $(|x| < 1, n \in \mathbb{R})$

Logarithms and Exponentials

$\log_a x = \frac{\log_b x}{\log_b a}$ $e^{x \ln a} = a^x$

Trigonometry

$\sin(A \pm B) = \sin A \cos B \pm \cos A \sin B$ $\cos(A \pm B) = \cos A \cos B \mp \sin A \sin B$

$\tan(A \pm B) = \frac{\tan A \pm \tan B}{1 \mp \tan A \tan B}$ $(A \pm B \neq (k + \frac{1}{2})\pi)$

$\sin A + \sin B = 2 \sin\frac{A+B}{2} \cos\frac{A-B}{2}$ $\sin A - \sin B = 2 \cos\frac{A+B}{2} \sin\frac{A-B}{2}$

$\cos A + \cos B = 2 \cos\frac{A+B}{2} \cos\frac{A-B}{2}$ $\cos A - \cos B = -2 \sin\frac{A+B}{2} \sin\frac{A-B}{2}$

Small Angle Approximations:

$\sin \theta \approx \theta$ $\cos \theta \approx 1 - \frac{\theta^2}{2}$ $\tan \theta \approx \theta$ where θ is measured in radians.

Differentiation

f(x)	f'(x)
$\tan kx$	$k \sec^2 kx$
$\sec kx$	$k \sec kx \tan kx$
$\cot kx$	$-k \csc^2 kx$
$\csc kx$	$-k \csc kx \cot kx$

First Principles: $f'(x) = \lim_{h \to 0} \frac{f(x + h) - f(x)}{h}$

For $y = \frac{f(x)}{g(x)}$, $\frac{dy}{dx} = \frac{f'(x)g(x) - f(x)g'(x)}{(g(x))^2}$

Integration (+ constant)

f(x)	$\int f(x)\,dx$
$\sec^2 kx$	$\frac{1}{k}\tan kx$
$\tan kx$	$\frac{1}{k}\ln|\sec kx|$
$\cot kx$	$\frac{1}{k}\ln|\sin kx|$

f(x)	$\int f(x)\,dx$
$\csc kx$	$-\frac{1}{k}\ln|\csc kx + \cot kx|$, $\frac{1}{k}\ln\left|\tan\left(\frac{1}{2}kx\right)\right|$
$\sec kx$	$\frac{1}{k}\ln|\sec kx + \tan kx|$, $\frac{1}{k}\ln\left|\tan\left(\frac{1}{2}kx + \frac{\pi}{4}\right)\right|$

$$\int u\frac{dv}{dx}\,dx = uv - \int v\frac{du}{dx}\,dx$$

Numerical Methods

The trapezium rule: $\int_a^b y\,dx \approx \frac{1}{2}h\{(y_0 + y_n) + 2(y_1 + y_2 + \ldots + y_{n-1})\}$, where $h = \frac{b-a}{n}$

The Newton-Raphson iteration for solving $f(x) = 0$: $x_{n+1} = x_n - \frac{f(x_n)}{f'(x_n)}$

Statistics

Probability

$P(A') = 1 - P(A)$ $P(A \cup B) = P(A) + P(B) - P(A \cap B)$

$P(A \cap B) = P(A)P(B|A)$ $P(A|B) = \frac{P(B|A)P(A)}{P(B|A)P(A) + P(B|A')P(A')}$

For independent events A and B:

$P(B|A) = P(B)$ $P(A|B) = P(A)$ $P(A \cap B) = P(A)P(B)$

Standard Deviation

Standard deviation = $\sqrt{\text{variance}}$ Interquartile range = IQR = $Q_3 - Q_1$

For a set of n values $x_1, x_2, \ldots x_i, \ldots x_n$:

$S_{xx} = \sum(x_i - \bar{x})^2 = \sum x_i^2 - \frac{(\sum x_i)^2}{n}$ Standard deviation = $\sqrt{\frac{S_{xx}}{n}}$ or $\sqrt{\frac{\sum x^2}{n} - \bar{x}^2}$

Discrete Distributions

Distribution of X	$P(X = x)$	Mean	Variance
Binomial $B(n, p)$	$\binom{n}{x}p^x(1-p)^{n-x}$	np	$np(1-p)$

Sampling Distributions

For a random sample of n observations from $X \sim N(\mu, \sigma^2)$: $\frac{\bar{X} - \mu}{\sigma/\sqrt{n}} \sim N(0, 1)$

Mechanics

Kinematics

For motion in a straight line with constant acceleration:

$v = u + at$ $s = ut + \frac{1}{2}at^2$ $s = \frac{1}{2}(u+v)t$ $s = vt - \frac{1}{2}at^2$ $v^2 = u^2 + 2as$

Statistical Tables

Percentage Points of the Normal Distribution

The z-values in the table are those which a random variable $Z \sim N(0, 1)$ exceeds with probability p, i.e. $P(Z > z) = 1 - \Phi(z) = p$.

p	z	p	z
0.5000	0.0000	0.0500	1.6449
0.4000	0.2533	0.0250	1.9600
0.3000	0.5244	0.0100	2.3263
0.2000	0.8416	0.0050	2.5758
0.1500	1.0364	0.0010	3.0902
0.1000	1.2816	0.0005	3.2905

Critical Values for Correlation Coefficients

This table concerns tests of the hypothesis that a population coefficient ρ is 0.
The values in the table are the minimum values which need to be reached by a sample correlation coefficient in order to be significant at the level shown, on a one-tailed test.

Sample Size	Product Moment Coefficient Level				
	0.10	0.05	0.025	0.01	0.005
4	0.8000	0.9000	0.9500	0.9800	0.9900
5	0.6870	0.8054	0.8783	0.9343	0.9587
6	0.6084	0.7293	0.8114	0.8822	0.9172
7	0.5509	0.6694	0.7545	0.8329	0.8745
8	0.5067	0.6215	0.7067	0.7887	0.8343
9	0.4716	0.5822	0.6664	0.7498	0.7977
10	0.4428	0.5494	0.6319	0.7155	0.7646
11	0.4187	0.5214	0.6021	0.6851	0.7348
12	0.3981	0.4973	0.5760	0.6581	0.7079
13	0.3802	0.4762	0.5529	0.6339	0.6835
14	0.3646	0.4575	0.5324	0.6120	0.6614
15	0.3507	0.4409	0.5140	0.5923	0.6411
16	0.3383	0.4259	0.4973	0.5742	0.6226
17	0.3271	0.4124	0.4821	0.5577	0.6055
18	0.3170	0.4000	0.4683	0.5425	0.5897
19	0.3077	0.3887	0.4555	0.5285	0.5751
20	0.2992	0.3783	0.4438	0.5155	0.5614
21	0.2914	0.3687	0.4329	0.5034	0.5487
22	0.2841	0.3598	0.4227	0.4921	0.5368
23	0.2774	0.3515	0.4133	0.4815	0.5256
24	0.2711	0.3438	0.4044	0.4716	0.5151
25	0.2653	0.3365	0.3961	0.4622	0.5052
26	0.2598	0.3297	0.3882	0.4534	0.4958
27	0.2546	0.3233	0.3809	0.4451	0.4869
28	0.2497	0.3172	0.3739	0.4372	0.4785
29	0.2451	0.3115	0.3673	0.4297	0.4705
30	0.2407	0.3061	0.3610	0.4226	0.4629
40	0.2070	0.2638	0.3120	0.3665	0.4026
50	0.1843	0.2353	0.2787	0.3281	0.3610
60	0.1678	0.2144	0.2542	0.2997	0.3301
70	0.1550	0.1982	0.2352	0.2776	0.3060
80	0.1448	0.1852	0.2199	0.2597	0.2864
90	0.1364	0.1745	0.2072	0.2449	0.2702
100	0.1292	0.1654	0.1966	0.2324	0.2565

The Binomial Cumulative Distribution Function

The values in the following tables show $P(X \leq x)$, where $X \sim B(n, p)$.

	$p =$	0.05	0.10	0.15	0.20	0.25	0.30	0.35	0.40	0.45	0.50
$n = 5$	$x = 0$	0.7738	0.5905	0.4437	0.3277	0.2373	0.1681	0.1160	0.0778	0.0503	0.0312
	1	0.9774	0.9185	0.8352	0.7373	0.6328	0.5282	0.4284	0.3370	0.2562	0.1875
	2	0.9988	0.9914	0.9734	0.9421	0.8965	0.8369	0.7648	0.6826	0.5931	0.5000
	3	1.0000	0.9995	0.9978	0.9933	0.9844	0.9692	0.9460	0.9130	0.8688	0.8125
	4	1.0000	1.0000	0.9999	0.9997	0.9990	0.9976	0.9947	0.9898	0.9815	0.9688
$n = 6$	$x = 0$	0.7351	0.5314	0.3771	0.2621	0.1780	0.1176	0.0754	0.0467	0.0277	0.0156
	1	0.9672	0.8857	0.7765	0.6554	0.5339	0.4202	0.3191	0.2333	0.1636	0.1094
	2	0.9978	0.9842	0.9527	0.9011	0.8306	0.7443	0.6471	0.5443	0.4415	0.3438
	3	0.9999	0.9987	0.9941	0.9830	0.9624	0.9295	0.8826	0.8208	0.7447	0.6563
	4	1.0000	0.9999	0.9996	0.9984	0.9954	0.9891	0.9777	0.9590	0.9308	0.8906
	5	1.0000	1.0000	1.0000	0.9999	0.9998	0.9993	0.9982	0.9959	0.9917	0.9844
$n = 7$	$x = 0$	0.6983	0.4783	0.3206	0.2097	0.1335	0.0824	0.0490	0.0280	0.0152	0.0078
	1	0.9556	0.8503	0.7166	0.5767	0.4449	0.3294	0.2338	0.1586	0.1024	0.0625
	2	0.9962	0.9743	0.9262	0.8520	0.7564	0.6471	0.5323	0.4199	0.3164	0.2266
	3	0.9998	0.9973	0.9879	0.9667	0.9294	0.8740	0.8002	0.7102	0.6083	0.5000
	4	1.0000	0.9998	0.9988	0.9953	0.9871	0.9712	0.9444	0.9037	0.8471	0.7734
	5	1.0000	1.0000	0.9999	0.9996	0.9987	0.9962	0.9910	0.9812	0.9643	0.9375
	6	1.0000	1.0000	1.0000	1.0000	0.9999	0.9998	0.9994	0.9984	0.9963	0.9922
$n = 8$	$x = 0$	0.6634	0.4305	0.2725	0.1678	0.1001	0.0576	0.0319	0.0168	0.0084	0.0039
	1	0.9428	0.8131	0.6572	0.5033	0.3671	0.2553	0.1691	0.1064	0.0632	0.0352
	2	0.9942	0.9619	0.8948	0.7969	0.6785	0.5518	0.4278	0.3154	0.2201	0.1445
	3	0.9996	0.9950	0.9786	0.9437	0.8862	0.8059	0.7064	0.5941	0.4770	0.3633
	4	1.0000	0.9996	0.9971	0.9896	0.9727	0.9420	0.8939	0.8263	0.7396	0.6367
	5	1.0000	1.0000	0.9998	0.9988	0.9958	0.9887	0.9747	0.9502	0.9115	0.8555
	6	1.0000	1.0000	1.0000	0.9999	0.9996	0.9987	0.9964	0.9915	0.9819	0.9648
	7	1.0000	1.0000	1.0000	1.0000	1.0000	0.9999	0.9998	0.9993	0.9983	0.9961
$n = 9$	$x = 0$	0.6302	0.3874	0.2316	0.1342	0.0751	0.0404	0.0207	0.0101	0.0046	0.0020
	1	0.9288	0.7748	0.5995	0.4362	0.3003	0.1960	0.1211	0.0705	0.0385	0.0195
	2	0.9916	0.9470	0.8591	0.7382	0.6007	0.4628	0.3373	0.2318	0.1495	0.0898
	3	0.9994	0.9917	0.9661	0.9144	0.8343	0.7297	0.6089	0.4826	0.3614	0.2539
	4	1.0000	0.9991	0.9944	0.9804	0.9511	0.9012	0.8283	0.7334	0.6214	0.5000
	5	1.0000	0.9999	0.9994	0.9969	0.9900	0.9747	0.9464	0.9006	0.8342	0.7461
	6	1.0000	1.0000	1.0000	0.9997	0.9987	0.9957	0.9888	0.9750	0.9502	0.9102
	7	1.0000	1.0000	1.0000	1.0000	0.9999	0.9996	0.9986	0.9962	0.9909	0.9805
	8	1.0000	1.0000	1.0000	1.0000	1.0000	1.0000	0.9999	0.9997	0.9992	0.9980
$n = 10$	$x = 0$	0.5987	0.3487	0.1969	0.1074	0.0563	0.0282	0.0135	0.0060	0.0025	0.0010
	1	0.9139	0.7361	0.5443	0.3758	0.2440	0.1493	0.0860	0.0464	0.0233	0.0107
	2	0.9885	0.9298	0.8202	0.6778	0.5256	0.3828	0.2616	0.1673	0.0996	0.0547
	3	0.9990	0.9872	0.9500	0.8791	0.7759	0.6496	0.5138	0.3823	0.2660	0.1719
	4	0.9999	0.9984	0.9901	0.9672	0.9219	0.8497	0.7515	0.6331	0.5044	0.3770
	5	1.0000	0.9999	0.9986	0.9936	0.9803	0.9527	0.9051	0.8338	0.7384	0.6230
	6	1.0000	1.0000	0.9999	0.9991	0.9965	0.9894	0.9740	0.9452	0.8980	0.8281
	7	1.0000	1.0000	1.0000	0.9999	0.9996	0.9984	0.9952	0.9877	0.9726	0.9453
	8	1.0000	1.0000	1.0000	1.0000	1.0000	0.9999	0.9995	0.9983	0.9955	0.9893
	9	1.0000	1.0000	1.0000	1.0000	1.0000	1.0000	1.0000	0.9999	0.9997	0.9990
$n = 12$	$x = 0$	0.5404	0.2824	0.1422	0.0687	0.0317	0.0138	0.0057	0.0022	0.0008	0.0002
	1	0.8816	0.6590	0.4435	0.2749	0.1584	0.0850	0.0424	0.0196	0.0083	0.0032
	2	0.9804	0.8891	0.7358	0.5583	0.3907	0.2528	0.1513	0.0834	0.0421	0.0193
	3	0.9978	0.9744	0.9078	0.7946	0.6488	0.4925	0.3467	0.2253	0.1345	0.0730
	4	0.9998	0.9957	0.9761	0.9274	0.8424	0.7237	0.5833	0.4382	0.3044	0.1938
	5	1.0000	0.9995	0.9954	0.9806	0.9456	0.8822	0.7873	0.6652	0.5269	0.3872
	6	1.0000	0.9999	0.9993	0.9961	0.9857	0.9614	0.9154	0.8418	0.7393	0.6128
	7	1.0000	1.0000	0.9999	0.9994	0.9972	0.9905	0.9745	0.9427	0.8883	0.8062
	8	1.0000	1.0000	1.0000	0.9999	0.9996	0.9983	0.9944	0.9847	0.9644	0.9270
	9	1.0000	1.0000	1.0000	1.0000	1.0000	0.9998	0.9992	0.9972	0.9921	0.9807
	10	1.0000	1.0000	1.0000	1.0000	1.0000	1.0000	0.9999	0.9997	0.9989	0.9968
	11	1.0000	1.0000	1.0000	1.0000	1.0000	1.0000	1.0000	1.0000	0.9999	0.9998

The binomial cumulative distribution function (continued)

	$p=$	0.05	0.10	0.15	0.20	0.25	0.30	0.35	0.40	0.45	0.50
$n=15$	$x=0$	0.4633	0.2059	0.0874	0.0352	0.0134	0.0047	0.0016	0.0005	0.0001	0.0000
	1	0.8290	0.5490	0.3186	0.1671	0.0802	0.0353	0.0142	0.0052	0.0017	0.0005
	2	0.9638	0.8159	0.6042	0.3980	0.2361	0.1268	0.0617	0.0271	0.0107	0.0037
	3	0.9945	0.9444	0.8227	0.6482	0.4613	0.2969	0.1727	0.0905	0.0424	0.0176
	4	0.9994	0.9873	0.9383	0.8358	0.6865	0.5155	0.3519	0.2173	0.1204	0.0592
	5	0.9999	0.9978	0.9832	0.9389	0.8516	0.7216	0.5643	0.4032	0.2608	0.1509
	6	1.0000	0.9997	0.9964	0.9819	0.9434	0.8689	0.7548	0.6098	0.4522	0.3036
	7	1.0000	1.0000	0.9994	0.9958	0.9827	0.9500	0.8868	0.7869	0.6535	0.5000
	8	1.0000	1.0000	0.9999	0.9992	0.9958	0.9848	0.9578	0.9050	0.8182	0.6964
	9	1.0000	1.0000	1.0000	0.9999	0.9992	0.9963	0.9876	0.9662	0.9231	0.8491
	10	1.0000	1.0000	1.0000	1.0000	0.9999	0.9993	0.9972	0.9907	0.9745	0.9408
	11	1.0000	1.0000	1.0000	1.0000	1.0000	0.9999	0.9995	0.9981	0.9937	0.9824
	12	1.0000	1.0000	1.0000	1.0000	1.0000	1.0000	0.9999	0.9997	0.9989	0.9963
	13	1.0000	1.0000	1.0000	1.0000	1.0000	1.0000	1.0000	1.0000	0.9999	0.9995
	14	1.0000	1.0000	1.0000	1.0000	1.0000	1.0000	1.0000	1.0000	1.0000	1.0000
$n=20$	$x=0$	0.3585	0.1216	0.0388	0.0115	0.0032	0.0008	0.0002	0.0000	0.0000	0.0000
	1	0.7358	0.3917	0.1756	0.0692	0.0243	0.0076	0.0021	0.0005	0.0001	0.0000
	2	0.9245	0.6769	0.4049	0.2061	0.0913	0.0355	0.0121	0.0036	0.0009	0.0002
	3	0.9841	0.8670	0.6477	0.4114	0.2252	0.1071	0.0444	0.0160	0.0049	0.0013
	4	0.9974	0.9568	0.8298	0.6296	0.4148	0.2375	0.1182	0.0510	0.0189	0.0059
	5	0.9997	0.9887	0.9327	0.8042	0.6172	0.4164	0.2454	0.1256	0.0553	0.0207
	6	1.0000	0.9976	0.9781	0.9133	0.7858	0.6080	0.4166	0.2500	0.1299	0.0577
	7	1.0000	0.9996	0.9941	0.9679	0.8982	0.7723	0.6010	0.4159	0.2520	0.1316
	8	1.0000	0.9999	0.9987	0.9900	0.9591	0.8867	0.7624	0.5956	0.4143	0.2517
	9	1.0000	1.0000	0.9998	0.9974	0.9861	0.9520	0.8782	0.7553	0.5914	0.4119
	10	1.0000	1.0000	1.0000	0.9994	0.9961	0.9829	0.9468	0.8725	0.7507	0.5881
	11	1.0000	1.0000	1.0000	0.9999	0.9991	0.9949	0.9804	0.9435	0.8692	0.7483
	12	1.0000	1.0000	1.0000	1.0000	0.9998	0.9987	0.9940	0.9790	0.9420	0.8684
	13	1.0000	1.0000	1.0000	1.0000	1.0000	0.9997	0.9985	0.9935	0.9786	0.9423
	14	1.0000	1.0000	1.0000	1.0000	1.0000	1.0000	0.9997	0.9984	0.9936	0.9793
	15	1.0000	1.0000	1.0000	1.0000	1.0000	1.0000	1.0000	0.9997	0.9985	0.9941
	16	1.0000	1.0000	1.0000	1.0000	1.0000	1.0000	1.0000	1.0000	0.9997	0.9987
	17	1.0000	1.0000	1.0000	1.0000	1.0000	1.0000	1.0000	1.0000	1.0000	0.9998
	18	1.0000	1.0000	1.0000	1.0000	1.0000	1.0000	1.0000	1.0000	1.0000	1.0000
$n=25$	$x=0$	0.2774	0.0718	0.0172	0.0038	0.0008	0.0001	0.0000	0.0000	0.0000	0.0000
	1	0.6424	0.2712	0.0931	0.0274	0.0070	0.0016	0.0003	0.0001	0.0000	0.0000
	2	0.8729	0.5371	0.2537	0.0982	0.0321	0.0090	0.0021	0.0004	0.0001	0.0000
	3	0.9659	0.7636	0.4711	0.2340	0.0962	0.0332	0.0097	0.0024	0.0005	0.0001
	4	0.9928	0.9020	0.6821	0.4207	0.2137	0.0905	0.0320	0.0095	0.0023	0.0005
	5	0.9988	0.9666	0.8385	0.6167	0.3783	0.1935	0.0826	0.0294	0.0086	0.0020
	6	0.9998	0.9905	0.9305	0.7800	0.5611	0.3407	0.1734	0.0736	0.0258	0.0073
	7	1.0000	0.9977	0.9745	0.8909	0.7265	0.5118	0.3061	0.1536	0.0639	0.0216
	8	1.0000	0.9995	0.9920	0.9532	0.8506	0.6769	0.4668	0.2735	0.1340	0.0539
	9	1.0000	0.9999	0.9979	0.9827	0.9287	0.8106	0.6303	0.4246	0.2424	0.1148
	10	1.0000	1.0000	0.9995	0.9944	0.9703	0.9022	0.7712	0.5858	0.3843	0.2122
	11	1.0000	1.0000	0.9999	0.9985	0.9893	0.9558	0.8746	0.7323	0.5426	0.3450
	12	1.0000	1.0000	1.0000	0.9996	0.9966	0.9825	0.9396	0.8462	0.6937	0.5000
	13	1.0000	1.0000	1.0000	0.9999	0.9991	0.9940	0.9745	0.9222	0.8173	0.6550
	14	1.0000	1.0000	1.0000	1.0000	0.9998	0.9982	0.9907	0.9656	0.9040	0.7878
	15	1.0000	1.0000	1.0000	1.0000	1.0000	0.9995	0.9971	0.9868	0.9560	0.8852
	16	1.0000	1.0000	1.0000	1.0000	1.0000	0.9999	0.9992	0.9957	0.9826	0.9461
	17	1.0000	1.0000	1.0000	1.0000	1.0000	1.0000	0.9998	0.9988	0.9942	0.9784
	18	1.0000	1.0000	1.0000	1.0000	1.0000	1.0000	1.0000	0.9997	0.9984	0.9927
	19	1.0000	1.0000	1.0000	1.0000	1.0000	1.0000	1.0000	0.9999	0.9996	0.9980
	20	1.0000	1.0000	1.0000	1.0000	1.0000	1.0000	1.0000	1.0000	0.9999	0.9995
	21	1.0000	1.0000	1.0000	1.0000	1.0000	1.0000	1.0000	1.0000	1.0000	0.9999
	22	1.0000	1.0000	1.0000	1.0000	1.0000	1.0000	1.0000	1.0000	1.0000	1.0000

The binomial cumulative distribution function (continued)

	$p=$	0.05	0.10	0.15	0.20	0.25	0.30	0.35	0.40	0.45	0.50
$n=30$	$x=0$	0.2146	0.0424	0.0076	0.0012	0.0002	0.0000	0.0000	0.0000	0.0000	0.0000
	1	0.5535	0.1837	0.0480	0.0105	0.0020	0.0003	0.0000	0.0000	0.0000	0.0000
	2	0.8122	0.4114	0.1514	0.0442	0.0106	0.0021	0.0003	0.0000	0.0000	0.0000
	3	0.9392	0.6474	0.3217	0.1227	0.0374	0.0093	0.0019	0.0003	0.0000	0.0000
	4	0.9844	0.8245	0.5245	0.2552	0.0979	0.0302	0.0075	0.0015	0.0002	0.0000
	5	0.9967	0.9268	0.7106	0.4275	0.2026	0.0766	0.0233	0.0057	0.0011	0.0002
	6	0.9994	0.9742	0.8474	0.6070	0.3481	0.1595	0.0586	0.0172	0.0040	0.0007
	7	0.9999	0.9922	0.9302	0.7608	0.5143	0.2814	0.1238	0.0435	0.0121	0.0026
	8	1.0000	0.9980	0.9722	0.8713	0.6736	0.4315	0.2247	0.0940	0.0312	0.0081
	9	1.0000	0.9995	0.9903	0.9389	0.8034	0.5888	0.3575	0.1763	0.0694	0.0214
	10	1.0000	0.9999	0.9971	0.9744	0.8943	0.7304	0.5078	0.2915	0.1350	0.0494
	11	1.0000	1.0000	0.9992	0.9905	0.9493	0.8407	0.6548	0.4311	0.2327	0.1002
	12	1.0000	1.0000	0.9998	0.9969	0.9784	0.9155	0.7802	0.5785	0.3592	0.1808
	13	1.0000	1.0000	1.0000	0.9991	0.9918	0.9599	0.8737	0.7145	0.5025	0.2923
	14	1.0000	1.0000	1.0000	0.9998	0.9973	0.9831	0.9348	0.8246	0.6448	0.4278
	15	1.0000	1.0000	1.0000	0.9999	0.9992	0.9936	0.9699	0.9029	0.7691	0.5722
	16	1.0000	1.0000	1.0000	1.0000	0.9998	0.9979	0.9876	0.9519	0.8644	0.7077
	17	1.0000	1.0000	1.0000	1.0000	0.9999	0.9994	0.9955	0.9788	0.9286	0.8192
	18	1.0000	1.0000	1.0000	1.0000	1.0000	0.9998	0.9986	0.9917	0.9666	0.8998
	19	1.0000	1.0000	1.0000	1.0000	1.0000	1.0000	0.9996	0.9971	0.9862	0.9506
	20	1.0000	1.0000	1.0000	1.0000	1.0000	1.0000	0.9999	0.9991	0.9950	0.9786
	21	1.0000	1.0000	1.0000	1.0000	1.0000	1.0000	1.0000	0.9998	0.9984	0.9919
	22	1.0000	1.0000	1.0000	1.0000	1.0000	1.0000	1.0000	1.0000	0.9996	0.9974
	23	1.0000	1.0000	1.0000	1.0000	1.0000	1.0000	1.0000	1.0000	0.9999	0.9993
	24	1.0000	1.0000	1.0000	1.0000	1.0000	1.0000	1.0000	1.0000	1.0000	0.9998
	25	1.0000	1.0000	1.0000	1.0000	1.0000	1.0000	1.0000	1.0000	1.0000	1.0000
$n=40$	$x=0$	0.1285	0.0148	0.0015	0.0001	0.0000	0.0000	0.0000	0.0000	0.0000	0.0000
	1	0.3991	0.0805	0.0121	0.0015	0.0001	0.0000	0.0000	0.0000	0.0000	0.0000
	2	0.6767	0.2228	0.0486	0.0079	0.0010	0.0001	0.0000	0.0000	0.0000	0.0000
	3	0.8619	0.4231	0.1302	0.0285	0.0047	0.0006	0.0001	0.0000	0.0000	0.0000
	4	0.9520	0.6290	0.2633	0.0759	0.0160	0.0026	0.0003	0.0000	0.0000	0.0000
	5	0.9861	0.7937	0.4325	0.1613	0.0433	0.0086	0.0013	0.0001	0.0000	0.0000
	6	0.9966	0.9005	0.6067	0.2859	0.0962	0.0238	0.0044	0.0006	0.0001	0.0000
	7	0.9993	0.9581	0.7559	0.4371	0.1820	0.0553	0.0124	0.0021	0.0002	0.0000
	8	0.9999	0.9845	0.8646	0.5931	0.2998	0.1110	0.0303	0.0061	0.0009	0.0001
	9	1.0000	0.9949	0.9328	0.7318	0.4395	0.1959	0.0644	0.0156	0.0027	0.0003
	10	1.0000	0.9985	0.9701	0.8392	0.5839	0.3087	0.1215	0.0352	0.0074	0.0011
	11	1.0000	0.9996	0.9880	0.9125	0.7151	0.4406	0.2053	0.0709	0.0179	0.0032
	12	1.0000	0.9999	0.9957	0.9568	0.8209	0.5772	0.3143	0.1285	0.0386	0.0083
	13	1.0000	1.0000	0.9986	0.9806	0.8968	0.7032	0.4408	0.2112	0.0751	0.0192
	14	1.0000	1.0000	0.9996	0.9921	0.9456	0.8074	0.5721	0.3174	0.1326	0.0403
	15	1.0000	1.0000	0.9999	0.9971	0.9738	0.8849	0.6946	0.4402	0.2142	0.0769
	16	1.0000	1.0000	1.0000	0.9990	0.9884	0.9367	0.7978	0.5681	0.3185	0.1341
	17	1.0000	1.0000	1.0000	0.9997	0.9953	0.9680	0.8761	0.6885	0.4391	0.2148
	18	1.0000	1.0000	1.0000	0.9999	0.9983	0.9852	0.9301	0.7911	0.5651	0.3179
	19	1.0000	1.0000	1.0000	1.0000	0.9994	0.9937	0.9637	0.8702	0.6844	0.4373
	20	1.0000	1.0000	1.0000	1.0000	0.9998	0.9976	0.9827	0.9256	0.7870	0.5627
	21	1.0000	1.0000	1.0000	1.0000	1.0000	0.9991	0.9925	0.9608	0.8669	0.6821
	22	1.0000	1.0000	1.0000	1.0000	1.0000	0.9997	0.9970	0.9811	0.9233	0.7852
	23	1.0000	1.0000	1.0000	1.0000	1.0000	0.9999	0.9989	0.9917	0.9595	0.8659
	24	1.0000	1.0000	1.0000	1.0000	1.0000	1.0000	0.9996	0.9966	0.9804	0.9231
	25	1.0000	1.0000	1.0000	1.0000	1.0000	1.0000	0.9999	0.9988	0.9914	0.9597
	26	1.0000	1.0000	1.0000	1.0000	1.0000	1.0000	1.0000	0.9996	0.9966	0.9808
	27	1.0000	1.0000	1.0000	1.0000	1.0000	1.0000	1.0000	0.9999	0.9988	0.9917
	28	1.0000	1.0000	1.0000	1.0000	1.0000	1.0000	1.0000	1.0000	0.9996	0.9968
	29	1.0000	1.0000	1.0000	1.0000	1.0000	1.0000	1.0000	1.0000	0.9999	0.9989
	30	1.0000	1.0000	1.0000	1.0000	1.0000	1.0000	1.0000	1.0000	1.0000	0.9997
	31	1.0000	1.0000	1.0000	1.0000	1.0000	1.0000	1.0000	1.0000	1.0000	0.9999
	32	1.0000	1.0000	1.0000	1.0000	1.0000	1.0000	1.0000	1.0000	1.0000	1.0000

The binomial cumulative distribution function (continued)

	$p =$	0.05	0.10	0.15	0.20	0.25	0.30	0.35	0.40	0.45	0.50
$n = 50$	$x = 0$	0.0769	0.0052	0.0003	0.0000	0.0000	0.0000	0.0000	0.0000	0.0000	0.0000
	1	0.2794	0.0338	0.0029	0.0002	0.0000	0.0000	0.0000	0.0000	0.0000	0.0000
	2	0.5405	0.1117	0.0142	0.0013	0.0001	0.0000	0.0000	0.0000	0.0000	0.0000
	3	0.7604	0.2503	0.0460	0.0057	0.0005	0.0000	0.0000	0.0000	0.0000	0.0000
	4	0.8964	0.4312	0.1121	0.0185	0.0021	0.0002	0.0000	0.0000	0.0000	0.0000
	5	0.9622	0.6161	0.2194	0.0480	0.0070	0.0007	0.0001	0.0000	0.0000	0.0000
	6	0.9882	0.7702	0.3613	0.1034	0.0194	0.0025	0.0002	0.0000	0.0000	0.0000
	7	0.9968	0.8779	0.5188	0.1904	0.0453	0.0073	0.0008	0.0001	0.0000	0.0000
	8	0.9992	0.9421	0.6681	0.3073	0.0916	0.0183	0.0025	0.0002	0.0000	0.0000
	9	0.9998	0.9755	0.7911	0.4437	0.1637	0.0402	0.0067	0.0008	0.0001	0.0000
	10	1.0000	0.9906	0.8801	0.5836	0.2622	0.0789	0.0160	0.0022	0.0002	0.0000
	11	1.0000	0.9968	0.9372	0.7107	0.3816	0.1390	0.0342	0.0057	0.0006	0.0000
	12	1.0000	0.9990	0.9699	0.8139	0.5110	0.2229	0.0661	0.0133	0.0018	0.0002
	13	1.0000	0.9997	0.9868	0.8894	0.6370	0.3279	0.1163	0.0280	0.0045	0.0005
	14	1.0000	0.9999	0.9947	0.9393	0.7481	0.4468	0.1878	0.0540	0.0104	0.0013
	15	1.0000	1.0000	0.9981	0.9692	0.8369	0.5692	0.2801	0.0955	0.0220	0.0033
	16	1.0000	1.0000	0.9993	0.9856	0.9017	0.6839	0.3889	0.1561	0.0427	0.0077
	17	1.0000	1.0000	0.9998	0.9937	0.9449	0.7822	0.5060	0.2369	0.0765	0.0164
	18	1.0000	1.0000	0.9999	0.9975	0.9713	0.8594	0.6216	0.3356	0.1273	0.0325
	19	1.0000	1.0000	1.0000	0.9991	0.9861	0.9152	0.7264	0.4465	0.1974	0.0595
	20	1.0000	1.0000	1.0000	0.9997	0.9937	0.9522	0.8139	0.5610	0.2862	0.1013
	21	1.0000	1.0000	1.0000	0.9999	0.9974	0.9749	0.8813	0.6701	0.3900	0.1611
	22	1.0000	1.0000	1.0000	1.0000	0.9990	0.9877	0.9290	0.7660	0.5019	0.2399
	23	1.0000	1.0000	1.0000	1.0000	0.9996	0.9944	0.9604	0.8438	0.6134	0.3359
	24	1.0000	1.0000	1.0000	1.0000	0.9999	0.9976	0.9793	0.9022	0.7160	0.4439
	25	1.0000	1.0000	1.0000	1.0000	1.0000	0.9991	0.9900	0.9427	0.8034	0.5561
	26	1.0000	1.0000	1.0000	1.0000	1.0000	0.9997	0.9955	0.9686	0.8721	0.6641
	27	1.0000	1.0000	1.0000	1.0000	1.0000	0.9999	0.9981	0.9840	0.9220	0.7601
	28	1.0000	1.0000	1.0000	1.0000	1.0000	1.0000	0.9993	0.9924	0.9556	0.8389
	29	1.0000	1.0000	1.0000	1.0000	1.0000	1.0000	0.9997	0.9966	0.9765	0.8987
	30	1.0000	1.0000	1.0000	1.0000	1.0000	1.0000	0.9999	0.9986	0.9884	0.9405
	31	1.0000	1.0000	1.0000	1.0000	1.0000	1.0000	1.0000	0.9995	0.9947	0.9675
	32	1.0000	1.0000	1.0000	1.0000	1.0000	1.0000	1.0000	0.9998	0.9978	0.9836
	33	1.0000	1.0000	1.0000	1.0000	1.0000	1.0000	1.0000	0.9999	0.9991	0.9923
	34	1.0000	1.0000	1.0000	1.0000	1.0000	1.0000	1.0000	1.0000	0.9997	0.9967
	35	1.0000	1.0000	1.0000	1.0000	1.0000	1.0000	1.0000	1.0000	0.9999	0.9987
	36	1.0000	1.0000	1.0000	1.0000	1.0000	1.0000	1.0000	1.0000	1.0000	0.9995
	37	1.0000	1.0000	1.0000	1.0000	1.0000	1.0000	1.0000	1.0000	1.0000	0.9998
	38	1.0000	1.0000	1.0000	1.0000	1.0000	1.0000	1.0000	1.0000	1.0000	1.0000